DESIGN OF SYSTEM ON A CHIP

Design of System on a Chip

Devices & Components

Edited by

Ricardo Reis

Universidade Federal do Rio Grande do Sul, Brasil

and

Jochen A.G. Jess

Eindhoven University of Technology, The Netherlands

KLUWER ACADEMIC PUBLISHERS

BOSTON / DORDRECHT / LONDON

A C.I.P. Catalogue record for this book is available from the Library of Congress.

ISBN 1-4020-7928-1 (HB)
ISBN 1-4020-7929-X (e-book)

Published by Kluwer Academic Publishers,
P.O. Box 17, 3300 AA Dordrecht, The Netherlands.

Sold and distributed in North, Central and South America
by Kluwer Academic Publishers,
101 Philip Drive, Norwell, MA 02061, U.S.A.

In all other countries, sold and distributed
by Kluwer Academic Publishers,
P.O. Box 322, 3300 AH Dordrecht, The Netherlands.

Printed on acid-free paper

Printed in the Netherlands.

Contents

Chapter 1

Design of Systems on a Chip: Introduction

[1]Ricardo Reis; [2]Jochen A. G. Jess
[1] Prof. at the Informatics Institute UFRGS – Federal Univ. of Rio Grande do Sul; P.O. Box 15064 – 91501-970 Porto Alegre, BRAZIL Tel: +55-51-316-6830, Fax: +55-51-3316-7308; E-mail: reis@inf.ufrgs.br

[2] Eindhoven University of Technology, p.o.box 513, 5600 MB Eidhoven, The Netherlands, Phone: 31-40-247-3353, Fax 31-40-246-4527

Key words: VLSI, microelectronics, roadmap, SoC.

Abstract: A short review of integrated circuit history is presented with a view in the effects of this revolution on the way of life. It goes on to say that Moore's law triggers a technology shockwave. To curb the entrepreneural risks the professional industry associations decided to anticipate the technology evolution by setting up roadmaps. The ITRS semiconductor roadmap was complemented by other roadmaps that preview the technology shockwave originating from the chip technology and propelling the product technology. The book content's focus is on devices and components for the design of systems on a chip. This chapter also presents an overview of the book contents.

1. MOORE'S LAW AND THE CONSEQUENCES

In 1947 John Bardeen, Walter Brattain and William Shockley invented the transistor. Except for perhaps a few experts the event went largely unnoticed. So had been the design of the world's first stored program computer, Konrad Zuse's Z3, completed in 1941. Nobody, not even the German military, was aware of the significance of this invention. At the same time, in Bletchley Park, in the UK, a team of dedicated people inspiringly guided by Alan Turing designed the "bomb". The bomb was a mechanical computing device, based on the ideas of the Polish mathematician Marjan Rejewski. Turing's version of it was able to break the code generated by the "Enigma" machine used by the German Navy. So the British Navy was able to decipher the messages of the German Navy, which controlled the movements of the German submarine fleet in the Atlantic. Therefore the allies succeeded to maneuver sufficient supplies across the Atlantic so as to prepare the invasion in Normandy, which essentially decided World War II in Europe. This fact remained largely unrecognized for almost three decades after the end of the war. All cryptographic activity was kept secret because of the Cold War situation emerging shortly after WW II was ended.

The bomb that brought scientific news on the public agenda was the nuclear bomb, the first of which was put to action on August 5, 1945. From that moment on scientific results became hot news items. But most people were interested in nuclear science exclusively because of the public perception that nuclear power would decide the next hot war. Only experts recognized the military potential of telecommunication and computers. Less than ten years after the invention of the transistor computers were built using them as essential switching elements. Jack Kilby from Texas Instruments created the first integrated circuit in 1958. Robert Noyce and Gordon Moore would establish companies like Fairchild and Intel. Another 13 years after the invention of the integrated circuit the first microprocessor, the Intel 4004, entered the market, carrying 2300 transistors on a single chip.

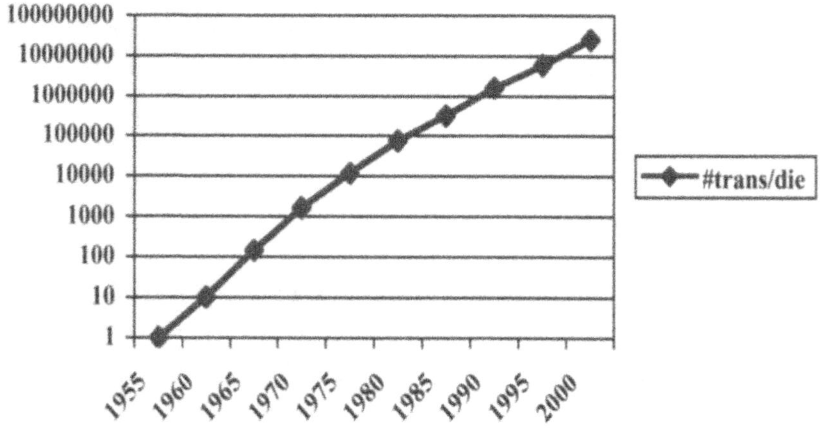

Figure 1. Moore's Law

Little by little the public became aware that there was a new technology advancing ever more prominently into the public domain. The automatic international telephone network set the first landmarks by connecting first the cities of one country, then the countries and eventually the continents. Computers started to penetrate from the scientific domain into the domain of financial transactions. Mass products were more and more manufactured by semi-automated production lines controlled by computers. But way before anybody ever recognized the significance of integrated circuits Gordon Moore realized the potential of them to establish a formidable economic phenomenon. Already in 1964, way before the appearance of the first microprocessor, he predicted an exponential growth of the density of switching functions on a single chip (see). Which means that he not only believed that it was technically feasible to control the complexity of very dense chips. His prediction implied that there would be financial support to build the necessary production lines and thus there would be a market of one or the other kind for chips of very high density.

But Gordon Moore was well ahead of the public. In the sixties the public mind was all occupied with space technology. In the summer of 1969 man landed on the moon as a result of the political efforts of the Kennedy and Johnson administrations. Rocket and nuclear technology paired up to establish a military threat that deeply penetrated into people's minds. Consequently even today people are emotionally opposed to nuclear energy to such an extent that the threats of a worldwide energy shortage and of the global warming phenomenon don't seem to count. Telecommunication was in the picture when the television frames with the moonwalkers illuminated the dusky living rooms all around the world. Simultaneously distorted voices uttered a specific idiom (from then on forever associated with flying of any kind) from which most people did not catch more than the continuously repeated phrase "Roger".

Moviemaker Stanley Kubrick had captured the doomsday sentiment of the public with respect to nuclear products adequately by creating Dr. Strangelove, a severely physically handicapped scientist and inventor of the "doomsday machine", the bomb that would end life on earth. In 1968, one year before the moon shot, and thus perfectly timed, he completed "2001 – A Space Odyssey" after a novel of Arthur C. Clarke. This movie captures the life in space quite adequately, so space scientists confirm even today. But it also reflects the public unawareness of the future face of information science. This is even more amazing as Kubrick attempted very seriously to anticipate the impact of supercomputing on areas like artificial intelligence. The drama develops within the space ship "Discovery", the brain of which is the supercomputer HAL. HAL represents a vision of ubiquitous intelligence: he runs the ship, he talks to the crew as a father, a friend or the boss that he actually is, depending on what he wants the crew to do and to feel. He is "Big Brother" and the long arm of the terrestrial space authority even to the point where he kills almost the entire crew because he thinks the crew is about to switch him off and to jeopardize the mission.

All these features go way beyond what artificial intelligence would ever prove to do. Today probably most serious artists would not engage into this kind of a vision. In a way it reflects the doomsday mentality of the cold war era. But it is really surprising that three years before the appearance of the first microprocessor on one chip there was no anticipation whatsoever how microelectronics would influence the interior of cockpits. Perhaps it is not so surprising that there are no laptops or palmtops in Discovery. Also nobody thought that display technology would change the presentation of data to become much more comprehensible. Similarly distributed computing and networking had not reached the artist's mind even though the ideas of computer networking were around and debated. The ARPA net, based on Paul Baran's and Donald Davies' idea of packet switching was about to become reality. The ARPA net used special purpose computers, so-called "interface message processors" (IMP), based on minicomputers, in this case the Honeywell H-516. The IMPs solved the problem to connect the vastly different so-called "hosts". Those hosts were the general-purpose computers local to the sites participating in the ARPA network project. The connections were actually established by leased telephone lines.

Instead HAL's brain is a compact piece of hardware arranged in a machine room with walls covered with a thick layer of printed boards. Obviously this layout was inspired by computers like the Remington Rand Univac 1 (were you could walk in through a door and feel like the brain's master), except that the tubes were replaced by transistors (see). A notion of time-sharing was all that entered into a piece of art supposed to render a serious vision of the far future. It proved to be outdated only some five years later.

Predictions are notoriously difficult. The preoccupation of his audience with the cold war fears and the relative lack of interest in telecommunications and computers can explain Kubrick's mistakes. Thirty years later things have become notably different. Electronics, computers, software, Internet, mobile telecommunications, embedded systems capture a great deal of attention of the public. All those items come under the label of "Information Technology". Stockbrokers invest and de-invest into it, students turn away from engineering in general except if the subject is related to it. Laymen handle the most sophisticated gadgets and children experience all states of joy handling "Play Stations" or "X-Boxes", which anytime in earlier history would have been addressed as supercomputers. It is not that predictions are any better now than they have been in the past. But

Moore's law and its various derivatives have been reasonably accurate for more than 35 years – through quite a number of economic crisis situations.

Figure 2. Remington Rand Univac 1 (1956); model on show in "Deutsches Museum München", Germany; **(Photo: J.A.G. Jess)**

2. THE "INTERNATIONAL ROADMAP FOR SEMICONDUCTOR TECHNOLOGY"

As the semiconductor fabrication technology evolved, the products based on it penetrated from the science into the military area, continuing through the regions of professionals like bankers, economists, managers and even attorneys and lawyers all the way into the range of consumers. The farther you go along this road the more erratic the market behavior becomes. Older industries serving the range of products from cars to detergents know all about that. The semiconductor production lines became more and more sophisticated. The business risk became larger and larger. Already in the early 90ies it cost about 1,5 Billion US$ to build a semiconductor fabrication line from scratch. In 1994 the US "Semiconductor Industry Association" (SIA) started an effort of "road mapping". The idea was to set the targets and the margins by associating process parameters like gate length, number of conducting layers or metal pitch with deadlines indicating when they were to be achieved. The industry hoped for a stabilization of the evolution to be able to curb the risk of investment. After only three years the roadmap from 1994 was outdated in many ways. It had unchained a fierce competition between the various market leaders (many of them in the Far East) which made all those targets look fairly conservative.

The initiative attracted a lot of attention. Today, next to the SIA, also four other associations sponsor the roadmap (which is now labeled as the "International Technology Roadmap for Semiconductors", ITRS): the "European Electronic Component Association" (EECA), the "Japan Electronics & Information Technology Industries Association" (JEITA), the "Korean Semiconductor Industry Association" (KSIA) and the "Taiwan Semiconductor Industry Association" (TSIA). International SEMATECH is the communication center for this activity. The roadmap document is essentially a large compendium of tables defining the evolution of technology parameters over the years. Paying tribute to the evolution of the various semiconductor products the current version of the roadmap has been thoroughly refined if compared to the 1994 SIA roadmap. The updating of the parameters in the roadmaps is an ongoing continuous process. To that end 15 "Technology Working Groups" (TWG) have been established meeting all year round to work on new numbers. The intermediate results are permanently available on the ITRS web site (http://public.itrs.net). An example of how the technology values have been updated over the years towards more aggressive values is illustrated in. While in 1994 the DRAM _ pitch in the years 2010 was predicted to become 70 nm this prediction was corrected to become 45 nm in the tables compiled in 2000.

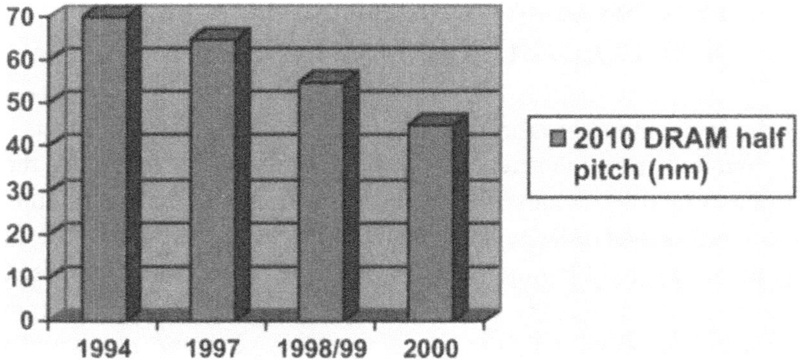

Figure 3. Predictions of the consecutive roadmaps for the DRAM half pitch for the year 2010

By way of an example we consider the predictions for DRAMs for 2014, which is the last year in the currently updated tables. The most optimistic scenario expects 48 Gbit DRAMs in production at a _ pitch of 30 nm on a chip of size 268 mm^2, yielding some 18,1 $Gbits/cm^2$. At the same time the introduction of 104 Gbit DRAMs is expected. While this size is based on the same _ pitch the chip size is expected to become 448 mm^2, which amounts to a density of 23,25 Gbits/cm^2. This is, by the way a downward correction if compared to the 1999 expectations. The 1999 tables predict a 194 Gbit DRAM on chip of size 792 mm^2. The progress of the roadmapping from 1999 to 2000 shows some more downward corrections even in the most optimistic scenarios. But in essence the characteristic growth of Moore's law is expected to stay intact till 2014.

3. THE "TECHNOLOGY SHOCKWAVE"

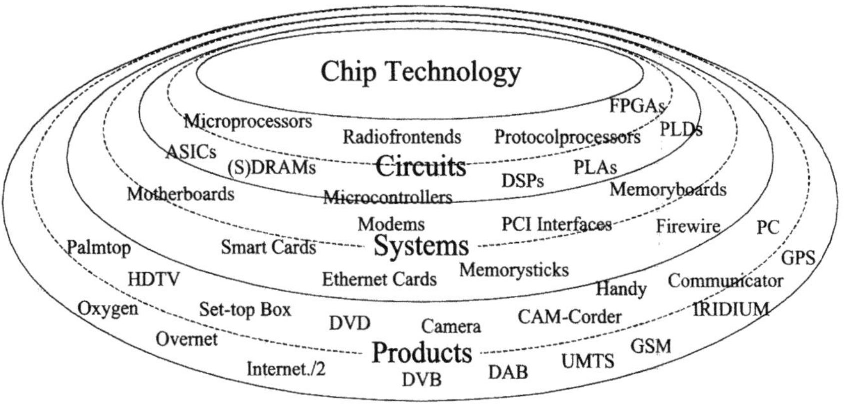

Figure 4. The "Si-Technology Shockwave"

The roadmap provided a tool of planning for all the industries depending on the chip industry. In the last thirty years the Si technology spawned a whole new industry making a large variety of new products. While those products enhanced the capability of almost anybody to compute and communicate in a way never anticipated, services existing already in the prewar period or in the fifties improved substantially. Telephone, radio, TV and all kinds of recording of sound and pictures have presented ever more new opportunities to the user. is supposed to give a visualization of the hardware products directly derived from the chip industry. In this visualization the Si technology resides at the epicenter of model. The chips currently on the market establish the first wave front of products. In the near future those chips will enter the market as so-called "Intellectual Property" (IP): chips will be so big that the available area cannot be utilized economically otherwise. Pieces of IP will have to be assembled on one chip into systems. This fact represents a formidable challenge to designers, the organization of design flows and the design automation industry.

The boards and cards we currently find in the gadgets and boxes we buy today make up for the next wave front in the model of . In the outermost wave front we see some of the consumer products and services available today as the consequence the chip technology.

4. THE TYDE OF THE MARKETS

It looks like the future exploitation of the chip technology will show even more shockwave phenomena. The availability of huge compounds of hardware spawned a blooming software industry. Above that we find communication to become one of the key issues. Communication on the chip will be one of the primary design issues in the near future. Software makers invented the "plug and play" concept, which is intended to have the non-experienced user connect hardware components and their associated software together easily. (The practitioner knows that it doesn't always work that way. You may insert a new interface card in one of your PCI slots and suddenly find your computer wouldn't shut down any more for unobvious reasons. Even an expensive helpdesk service wouldn't relieve you from the experience of feeling like a dumb and underprivileged individual. But all of us appreciate the idea!).

Another item stirring the public was the breakthrough in mobile communication in the nineties. Of course the military was using mobile communication already in WW II. But even with the advent of semiconductors mobile communication was restricted to the realm of professional systems. In Europe the use of mobile communication spread from sparsely populated but technologically highly developed areas. Those qualifications apply in particular to Scandinavia, but also to countries of the Southern hemisphere like Brazil. Up in the North of Europe people's life would often enough depend on a radio link. No wonder that companies like the Swedish Ericsson and the Finnish Nokia achieved a major market position. (Nokia, by the way, started with making fisherman's supplies such as rubber boots – but of course fishermen, too, needed a lot of radios traditionally!)

In putting the concept of the "World Wide Web" on top of the existing global and local computer network infrastructure Tim Berners-Lee and Robert Cailliau set out to create "a pool of human knowledge" (1994). They realized that the essence of knowledge (as compared to sheer data) is the ability to link contents together regardless of where they physically reside. This is the basic idea of the "Hypertext Mark-up Language" (HTML) enabling everybody to assemble websites from locally distributed data. This way there arises a web of contents, where the physical links are no longer visible (and relevant) and are replaced by conceptual links, which are supposed "to make sense". Together with powerful browsers, servers, routers and a versatile mail facility (enhanced by the "attachment" option) a new world has been opened spawning a wealth of business activity. Today the makers of consumer products, mobile phones and the computer industry are engaged in a fierce competition for a major share in web technology.

It is not surprising that the simultaneous appearance of mobile phones and web services on the market created a major shockwave by itself. To begin with the social opposition to the technological innovation was low. While problems like the energy shortage, global warming and traffic congestion created powerful oppositional activities the complaints against information technology touched issues like the possible radiation damage by the use of mobile phones or the density of antennas for mobile communication on buildings. Also there was the traditional criticism on the content of the media (notably television) and the fear regarding the disruption of social structures by the overuse of communication media. But all this did not coagulate to a movement powerful enough to put a halt to the enthusiasm of the public when adapting the new media. The public resistance that belongs to the daily grief of managers in the nuclear and chemical industry and the board chairmen of the car and airplane manufacturers (and that's not to mention the managers of airports!) was almost totally absent when it came to Internet and mobile phones. Indeed, many of those technologies were deemed capable of resolving some of the mobility and congestion problems we experience every day. Looking at the market developments the term "new economy" reached the newspaper columns and talk show presentations, denoting the combined phenomenon of steep economic growth and low inflation rates (at least in the US and Europe!).

In the meantime (in the summer of 2001) we seem to be back to the old economy again. The year 2001 definitely stopped the boom of the information technology. In Central Europe and the US inflation is back on the agenda. The growth of Internet use stagnates. The sales in computers and mobile equipment decrease spectacularly. The transition to the third generation of mobile phone service, the so-called "Universal Mobile Telecommunication Service" (UMTS), may have to be postponed for several years. This undermines the financial position of a number of European telecommunication service providers, who had to acquire sizable loans to buy the licenses for the appropriate frequency bands and to prepare for the huge investments in new technical infrastructure. Those phenomena backfire on the chipmakers. Sales of chips have been down by thirty percent or more recently. Fabrication lines run on half of their usual load. Consequently large orders for chip manufacturing equipment have been cancelled.

As if all this wasn't enough this went along with a major collapse of the stock market. It started with a major shakeout between Internet providers and servers for "Electronic Commerce" with insufficiently stable business models. It then reached out for the telecommunication providers. The shares of some of those lost 90% of their value within a few months. Thus there evaporated the potential to finance new infrastructure by issuing new shares.

Is this the end of information technology? Is roadmapping a pointless exercise from now on? It is hard to believe. But there is no doubt that growth rates such as those of the most recent years will not come back for some time. Eventually the roadmap may experience some delay. This delay is not the result of physical limitations or our inability to install the technology. Rather the market will impose its pace of acceptance of the new products and services. Yet the potential of the Silicon technology is far from exhausted. More than that: there is a growing need of products and services for communication in view of the limits of mobility end energy in order to maintain the world trade. But it may be necessary to pay more attention to the voice of the market. Rather than just putting down a roadmap for the technology, coordinated planning between technologists, product makers and service providers may be necessary to control the business risk. For instance the total infrastructure of optical fiber backbones is reported to exhibit an overcapacity of two to three orders of magnitude. The rates for international calls (or even intercontinental calls) are in the same range as those for local calls. On the other hand the bandwidth limitations in the residential subscriber loop are still impairing the use of Internet for the common user. On one hand DSL and ADSL are expensive for such a user. On the other hand the common user is likely to have requests needing a lot of bandwidth. While he can acquire a digital camcorder for a reasonable price he hardly can afford to mail even small pieces of his videos to his friends and relatives as an attachment to a mail message. Also the downloading or display of video content via Internet meets with serious bandwidth limitations.

If investment is scarce it may be worthwhile to complement the semiconductor roadmap with service and bandwidth roadmaps. The results of such a planning activity may guide investments to more long-term profit to the benefit of everybody. The gold rush phenomenon of the late nineties may prove too wasteful and may destroy the investment into many years of research.

5. THE FIRST BOOK: SEMICONDUCTOR DEVICES AND COMPONENTS

This book is the first of two volumes addressing the design challenges associated with new generations of the semiconductor technology. The subjects deal with issues closely related to the epicenter of. The various chapters are the compilations of tutorials presented at workshops in Brazil in the recent years by prominent authors from all over the world. In particular the first book deals with components and circuits. To begin with device models have to satisfy the conditions to be computationally economical in addition to being accurate and to scale over various generations of technology. Colin McAndrew's paper addresses bipolar transistors while Matthias Bucher and Christian Enz deal with MOS transistor models.

An important problem is that of statistical variations of process parameters. Those variations translate into variations of circuit behavior which are directly related to the so-called "parametric yield loss" associated with the mass production of chips. In a second contribution by Colin McAndrew we learn about how to deal with statistical variations when performing circuit simulation. This is a matter of computational efficiency and sound physical analysis and eventually may decide about the issue of "design for manufacturability".

The next level of complexity is that of circuit components. The fast transition between consecutive generations of technology and causes the complete redesign of circuits for every new technology generation to be uneconomical. Therefore José Franca discusses an approach to generate blocks like data converters, amplifiers and filters and assemble them to form systems matching a range of applications and technologies. The four main ingredients of such a methodology are optimized system level partitioning, technology adaptation by appropriate component design, efficient Silicon area use by sophisticated area planning techniques and finally advanced wire analysis and route planning.

While it obvious that the main advances in technology are associated with the shrinking of the lateral pitches of transistors and wires technologists decided that the small features could actually only put to use if the signal levels would be scaled down. So from 1994 onwards the standard supply voltage of 5 Volt was replaced by 3 Volts. The further progress of Silicon technology shows a continued decrease of supply voltage levels all the way to 0,9 Volt. This is the essential measure to control power dissipation in the Silicon structures. Yet the threat is in contaminating phenomena that don't scale with the supply voltage such as for instance random parameter variations. A group of authors headed by M. Ismail deals with low power low voltage square law CMOS composite transistors and design techniques ensuring robust low power analog circuits. In the same line of thought Kiyoo Itoh discusses the DRAM and SRAM cells for the range between 0,5 Volt to 2 Volt. Again parameter variations are a key issue along with subthreshold current suppression. The paper also turns to "Silicon on Insulator" (SOI) solutions. Eventually R. Leung approaches the issue of input-output buffers (I/O buffers) with an emphasis on low voltage differential signaling buffers.

Winding up this first book is a contribution by Takayuki Yanagawa and Sergio Bampi. They discuss the background of the ITRS roadmap. The roadmap simply states the value of key features of the technology but it does not tell what it takes to have those features available for stable mass production. Yanagawa and Bampi expose the problems that have to be solved and the main lines of research which have to be pursued.

6. REFERENCES

[HODG83] Hodges, A., (1983) Alan Turing: the Enigma, Walker & Co., New York.

[SIN99] Singh, S., (1999) The Code Book, Fourth Estate Ltd., London.

[TRAN93] Frank, F.C., (Editor), (1993) Operation Epsilon: the Farm Hall Transcripts, Institute of Physics Publishing, Bristol and Philadelphia.

[CLA68] Clarke, A.C., (1968) 2001 – A Space Odyssey, Little, Brown & Co., London.

[STO97] Stork, D.G. (Editor), (1997) HAL's Legacy: 2001's Computer as Dream and Reality, The MIT Press, Cambridge, Mass. and London.

[ABB99] Abbate, J., (1999) Inventing the Internet, The MIT Press, Cambridge, Mass. and London.

[BER99] Berners-Lee, T., with Fischetti, M., (1999) Weaving the Web, HarperCollins Publishers Inc., New York.

[ITRS] The International Technology Roadmap for Semiconductors, http://public.itrs.net

Chapter 2

BJT Modeling with VBIC

C. C. McAndrew
Motorola, Inc., 2100 East Elliot Road, Tempe AZ, 85284 U.S.A., PH:(602)413-3982
FAX:(602)413-5343, mcandrew@sst.sps.mot.com

Key words: VBIC, bipolar transistor modeling, Gummel-Poon model, SPICE modeling, compact modeling, electrothermal modeling, self heating

Abstract: The SPICE Gummel-Poon model has served the IC industry well, however it is not sufficiently accurate for design in modern bipolar and BiCMOS technologies. This tutorial reviews the VBIC model, and highlights its main features: improved Earlyeffect modeling, parasitic substrate transistor modeling, quasi-saturation modeling, improved temperature modeling, impact ionization modeling, and electrothermal modeling.

1. INTRODUCTION

For over 20 years the SPICE Gummel-Poon (SGP) model (Gummel, 1970; Nagel, 1975) has been the IC industry standard for circuit simulation for bipolar junction transistors (BJTs). This is a testament to the sound physical basis of the model.However, the SGP model is not perfect. Some of the shortcomings of the SGP modelhave been known for a long time, such as its inability to model collector resistancemodulation (quasi-saturation) and parasitic substrate transistor action. And theinexorable advance of IC manufacturing technologies has magnified theinaccuracies in other aspects of the SGP model, e.g. the Early effect formulation for modeling output conductance .

Improved BJT models have been presented (Turgeon, 1980; Kull, 1985; de Graaff, 1985; Stubing, 1987; Jeong, 1989), however none have become an industry standard to replace the SGP model. VBIC was defined by a group of representatives from the IC and CAD industries to try to rectify this situation. VBIC is public domain, and complete source code is publicly available. VBIC is also as similar as possible to the SGP model, to leverage the existing knowledge and training of characterization and IC design engineers.

The following are the main modeling enhancements of VBIC over SGP:

- improved Early effect ( ) modeling
- quasi-saturation modeling
- parasitic substrate transistor modeling
- parasitic fixed (oxide) capacitance modeling
- avalanche multiplication modeling
- improved temperature dependence modeling
- decoupling of base and collector currents
- electrothermal (self heating) modeling
- C_∞ continuous (smooth) modeling
- improved heterojunction bipolar transistor (HBT) modeling.

The additional capabilities of VBIC are turned off with the default values of its model parameters, so VBIC defaults to being close to the SGP model, the exception being the Early effect model which is different between the two models. The presentation and examples used here are for 4-terminal vertical NPN transistors. VBIC can also be used for vertical PNP modeling, and for HBT modeling, but it is not directly targeted at lateral BJT modeling. Vertical PNPs in smartpower technologies are often 5-terminal devices, and VBIC can be used in a subcircuit to model such devices, however this does not properly model transistor action of the second parasitic BJT.

Compact models for circuit simulation should scale properly with device geometry. However, for BJTs the plethora of layout topologies and structure make this impossible to do in a comprehensive manner. Therefore VBIC explicitly does not include any geometry mappings. It is assumed that geometry scaling for VBIC will be handled either in pre-processing for the generation of model libraries for circuit simulation, or via scaling relations specific to a particular technology implemented either in the simulator or the CAD system used for design.

2. VBIC EQUIVALENT NETWORK

Figure 1 shows the equivalent network of VBIC, which includes an intrinsic NPN transistor, a parasitic PNP transistor, parasitic resistances and capacitances, a local thermal network (used only with the electrothermal version of the model), and a circuit that implements excess phase for the forward transport current I_{tzf}.

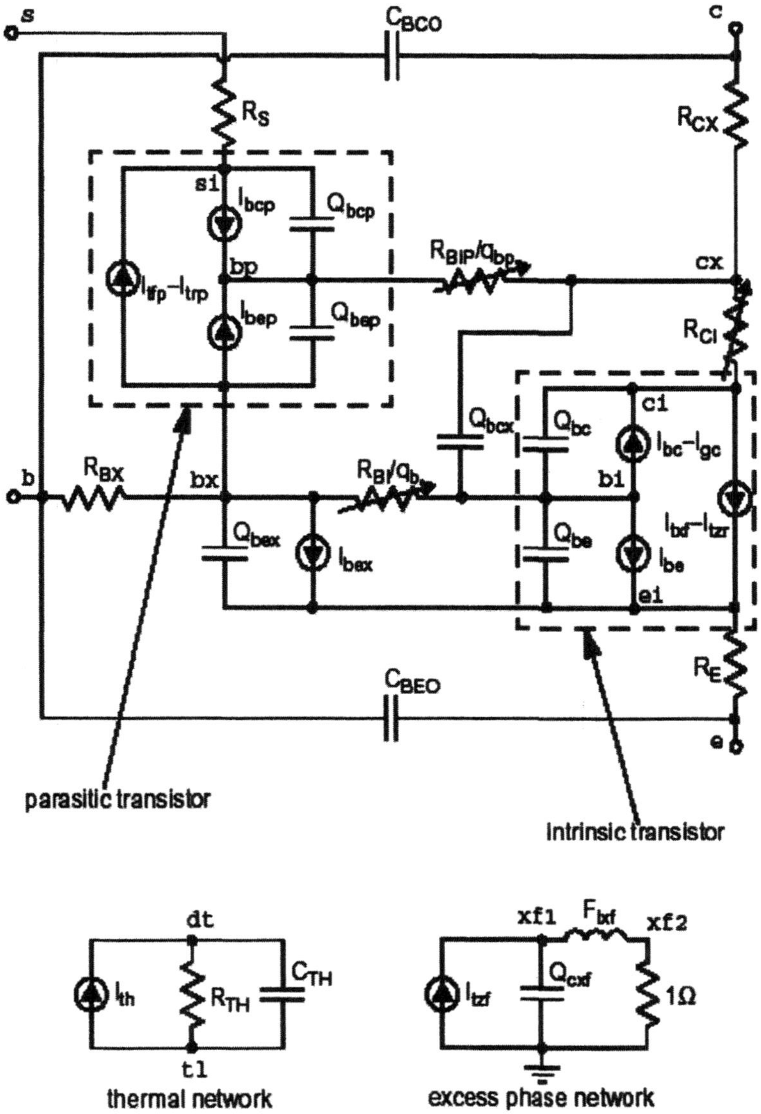

Figure 1. VBIC equivalent network

For the electrothermal version of VBIC the branch currents and charges in the electrical part of the model also depend on the local temperature rise, the voltage on the node dt. The thermal equivalent circuit includes two nodes external to the model so that the local heating and dissipation can be connected to a thermal network that models the thermal properties of the material in which the BJT and surrounding devices are built.

Table 1 lists the elements of the VBIC equivalent network.

Table 1. Elements of VBIC equivalent network

Name	*Element*
I_{tzf}	forward transport current, zero phase
I_{t0xf}	forward transport current, with excess phase
$Q_{cxf} F_{lxf}$	excess phase circuit capacitance and inductance
I_{tzr}	reverse transport current, zero phase
I_{be}	intrinsic base-emitter current
I_{bex}	extrinsic (side) base-emitter current
Q_{be}	intrinsic base-emitter charge (depletion and diffusion)
Q_{bex}	extrinsic (side) base-emitter charge (depletion only)
I_{bc}	intrinsic base-collector current
I_{gc}	base-collector weak avalanche current
Q_{bc}	intrinsic base-collector charge (depletion and diffusion)
Q_{bcx}	extrinsic base-collector charge (diffusion only)
I_{ccp}	parasitic transistor transport current
I_{bep}	parasitic base-emitter current
Q_{bep}	parasitic base-emitter charge (depletion and diffusion)
I_{bcp}	parasitic base-collector current

3. VBIC MODEL FORMULATION

The core of VBIC, as with most BJT models, is the transport (collector) currentmodel, which follows directly from Gummel (1970). For electrons, the continuity equation is

$$\nabla \bullet J_e = q\left(\frac{\partial n}{\partial t} + R_e - G_e\right) \tag{1}$$

where J_e is the electron current density, is the q magnitude of the electronic charge, n is the electron concentration, and R_e and G_e are the electron recombination and generation rates, respectively. The drift-diffusion relation for electrons is

$$J_e = \mu_e(kT\nabla n - qn\nabla\psi) = -q\mu_e n\nabla\phi_e \tag{2}$$

where μ_e is the electron mobility, k is Boltzmann's constant, T is the temperature in degrees Kelvin, ψ is the electrostatic potential, and ϕ_e is the electron quasi-Fermi potential. The electron concentration is

$$n = n_{ie}\exp\left(\frac{\psi - \phi_e}{V_{tv}}\right) \tag{3}$$

where n_{te} is the effective intrinsic concentration, including bandgap narrowing, and $V_{tv} = kT/q$ is the thermal voltage, and

$$p = n_{ie}\exp\left(\frac{\phi_h - \psi}{V_{tv}}\right) \tag{4}$$

gives the hole concentration p, where ϕ_h is the hole quasi-Fermi potential.

Analysis of the transport in the base region of a BJT is based on equations (1) and (2). In the steady state in the x dimension only, ignoring recombination and generation (which is generally reasonable for the base of a BJT), gives

$$\frac{\partial J_{ex}}{\partial x} = 0 \tag{5}$$

$$J_{ex} = \text{constant} = -q\mu_e n\frac{\partial\phi_e}{\partial x} = q\mu_e n_{ie} V_{tv}\exp\left(\frac{\psi}{V_{tv}}\right)\frac{\partial \exp(-\phi_e/V_{tv})}{\partial x} \tag{6}$$

$$J_{ex} = qV_{tv}\frac{\exp(-\phi_{ew}/V_{tv}) - \exp(-\phi_{e0}/V_{tv})}{\int_0^w \frac{\exp(-\psi/V_{tv})}{\mu_e n_{ie}}dx} \tag{7}$$

directly from equation (1), hence
follows after some manipulation. Integrating equation (6) from the emitter ($x = 0$) to the collector ($x = w$) through the base gives

where ψ, μ_e and n_{ie} are all functions of position x. Multiplying both the numerator and the denominator of equation (7) by $\exp(\phi_h/V_{tv})$, and noting that the difference between hole and electron quasi-Fermi potentials across a junction is just the voltage applied across the junction, gives

$$J_{ex} = qV_{tv}\frac{\exp(V_{bci}/V_{tv}) - \exp(V_{bei}/V_{tv})}{\int_0^w \frac{p}{\mu_e n_{ie}^2}dx} \tag{8}$$

where V_{bei} is the intrinsic base-emitter voltage, between nodes `bi` and `ei` of the equivalent network of Figure 1, and V_{bci} is the intrinsic base-collector voltage, between nodes `bi` and `ci`. Equation (8) is the basis of Gummel-Poon type BJT models (Gummel 1970). It shows that the collector current varies exponentially with applied bias, and is controlled by the integrated base charge, which is commonly called the base Gummel number.

Use of equation (8) in VBIC requires the base charge to be modeled as a function of applied bias. For VBIC the base charge is normalized with respect to its value at zero applied bias, and includes depletion and diffusion components (Gummel, 1970; Getreu, 1976). For VBIC the forward and reverse transport currents are

$$I_{tzf} = I_S\frac{\exp(V_{bei}/(N_F V_{tv})) - 1}{q_b} \tag{9}$$

$$I_{tzr} = I_S\frac{\exp(V_{bci}/(N_R V_{tv})) - 1}{q_b} \tag{10}$$

where I_S is the transport saturation current, N_F and N_R are the forward and reverse ideality factors, and q_b is the normalized base charge. The ideality factors are introduced as parameters, rather being forced to be 1, to allow flexibility in modeling, to recognize that the theoretical analyses above are approximate, because non-ideal transport behavior is observed in HBTs, and for compatibility with the SGP models.

It can be shown that under the restrictive assumption that each branch of the equivalent network for VBIC must be passive (rather than the general condition that the whole model must be passive), then the conditions $N_R \geq N_F$ $N_F \geq N_R$ and must hold, the only solution being $N_R \cong N_F$. It is recommended that for silicon devices this equality is maintained. For HBTs this restriction can be relaxed.

The normalized base charge is

$$q_b = q_1 + \frac{q_2}{q_b}$$
$$q_1 = 1 + \frac{q_{je}}{V_{ER}} + \frac{q_{jc}}{V_{EF}} \qquad (11)$$
$$q_2 = \frac{I_S(\exp(V_{bei}/(N_F V_{tv})) - 1)}{I_{KF}} + \frac{I_S(\exp(V_{bci}/(N_R V_{tv})) - 1)}{I_{KR}}$$

where V_{EF} and V_{ER} are the forward and reverse Early voltages I_{KF} and I_{KR} and are the forward and reverse knee currents. The normalized depletion charges are

$$q_{je} = q_j(V_{bei}, P_E, M_E, F_C, A_{JE}), q_{jc} = q_j(V_{bci}, P_C, M_C, F_C, A_{JC}) \qquad (12)$$

where P_E and P_C are the built-in potentials M_E and M_C and are the grading coefficients of the base-emitter and base-collector junctions, respectively. The normalized depletion charge function q_j is such that

$$c_j(V, P, M) = \frac{\partial q_j(V, P, M)}{\partial V} \approx \frac{1}{(1 - V/P)^M} \qquad (13)$$

for reverse and low forward bias, and if the depletion capacitance smoothing parameters A_{JE} and A_{JC} are less than zero c_j smoothly limits to its value at $F_C P$, else c_j linearly increases for $V > F_C P$ to match the SGP model, see Figure 2.

The Early voltage components model the variation in q_b caused by changes in the depletion regions at the base-emitter and base-collector junctions, and the knee current components model the effects of high level injection. In this analysis the high level injection is considered to be in the base, whereas in normal NPNs it occurs when the base pushes out into the more lightly doped collector. This is handled in VBIC with the quasi-saturation model detailed below.

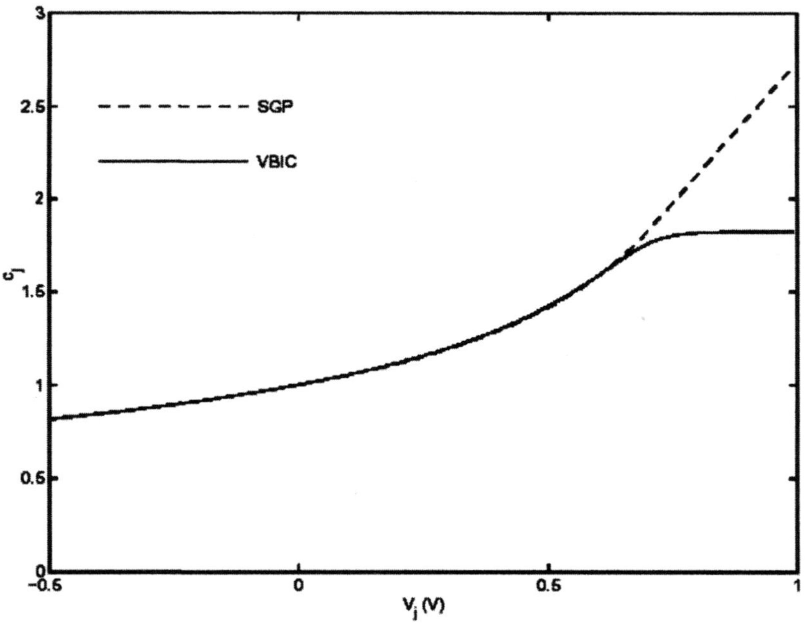

Figure 2. C_∞ continuous normalized capacitance model

If the excess phase delay T_D is set to zero then I_{txf} in Figure 1 is just the I_{tzf} of equation (9). If $T_D > 0$ then the capacitance and inductance of the excess phase network of Figure 1 are set to T_D and $T_D / 3$ respectively, and in the *s* domain the transfer function of the excess phase network is

$$V(\mathtt{xf2}) = \frac{3I_{tzf}/T_D^2}{s^2 + 3s/T_D + 3/T_D^2} \tag{14}$$

which implements a second order polynomial approximation to ideal excess phase (Weil, 1978). The voltage on node `xf2` is then directly used as I_{txf}. This implementation of excess phase is consistent between small signal and transient analyses in a circuit simulator, and is independent of the numerical algorithms within a simulator. This is an advantage over an ideal excess phase model, as used in some simulators, which can only be implemented for small signal analysis and therefore leads to inconsistencies between small signal and transient simulations. Implementation of a direct form of equation (14) depends on the numerical integration algorithms employed (Weil, 1978), whereas the equivalent network approach does not. Note that although the excess phase network in Figure 1 looks like it introduces 3 extra unknowns (2 node voltages and the inductor current) into the modified nodal formulation commonly used within circuit simulators, it can actually be implemented with only 2 additional simulation variables, by taking advantage of the observation that the voltage at node `xf2` is equivalent to the current in the inductor.

The intrinsic charges are

$$Q_{be} = C_{JE}W_{BE}q_{je} + \tau_F I_{tzf} \tag{15}$$

where C_{JE} is the zero bias base-emitter depletion capacitance, W_{BE} is the partitioning of the base-emitter depletion capacitance between intrinsic and extrinsic components, q_{je} is defined in equation (12), and τ_F is the forward transit time, modeled as

$$\tau_F = T_F(1 + Q_{TF}q_1)\left(1 + X_{TF}\exp\left(\frac{V_{bci}}{1.44V_{TF}}\right)\left(\frac{q_b I_{tzf}}{q_b I_{tzf} + I_{TF}}\right)^2\right) \tag{16}$$

which is the SGP model, with an additional term in q_1 added to model the change in base transit time as the base-emitter and base-collector depletion region edges move with bias. The extrinsic base-emitter is charge is

$$Q_{bex} = C_{JE}(1 - W_{BE})q_j(V_{bex}, P_E, M_E, F_C, A_{JE}) \tag{17}$$

where it is apparent that $0 \leq W_{BE} \leq 1$ should hold, and V_{bex} is the extrinsic caseemitter bias, between nodes `bx` and `ei` in Figure 1.

The intrinsic base-collector is

$$Q_{bc} = C_{JC}q_{jc} + T_R I_{tzr} + Q_{CO}K_{bci} \tag{18}$$

where C_{JC} is the zero bias base-collector depletion capacitance, q_{jc} is defined in equation (12), and T_R is the reverse transit time. The term $Q_{CO}K_{bci}$ models the diffusion charge associated with base pushout into the collector, and it and a similar extrinsic term

$$Q_{bcx} = Q_{CO}K_{bcx} \tag{19}$$

will be addressed below.

The base charge appears in both the transport current model, via the normalized base charge, and explicitly in the charge elements. By comparing the equivalent terms it is apparent that

$$C_{JE}V_{ER} = C_{JC}V_{EF} = T_F I_{KF} = T_R I_{KR} = \int_0^w p\,dx = Q_{b0} \tag{20}$$

should be true for the transport and charge models to be consistent (where the secondterm should only include the portion of C_{JC} under the emitter).

This is not enforced in VBIC, for compatibility with SGP, and to allow more degrees of freedom in fitting measured device characteristics.

The major difference between the above transport current formulation of VBIC and that of SGP is the Early effect modeling via the q_1 term. In SGP this is approximated by (McAndrew, 1996)

$$q_1 = \frac{1}{1 - \frac{V_{bei}}{V_{AR}} - \frac{V_{bci}}{V_{AF}}}. \tag{21}$$

Equation (21) cannot model the bias dependence of output conductance well over a wide range of biases, because it has linearized the dependence of depletion charge on applied bias. Figure 3 compares I_e / g^r_o modeling of VBIC and SGP. SGP cannot even qualitatively model the observed trends in measured data, it has the linear variation of equation (21) whereas VBIC models the onset of a fully depleted base region well. Therefore for improved accuracy of modeling, backward compatibility of VBIC to SGP for the Early effect modeling was not maintained.

The base current elements of VBIC model recombination and generation currents. Three mechanisms are important, Shockley-Read-Hall recombination,

$$R_{srh} = \frac{np - n_{ie}^2}{\tau_h(n + n_{ie}) + \tau_e(p + n_{ie})}, \tag{22}$$

where τ_e and τ_h are the electron and hole trapping lifetimes, respectively, Auger recombination,

$$R_{aug} = (c_e n + c_h p)(np - n_{ie}^2), \tag{23}$$

where c_e and c_h are the Auger rate constants for electrons and holes, respectively, and surface hole recombination, modeled as a recombination current density

$$J_{hec} = qS_h(p - p_{ec0}) \tag{24}$$

where S_h is the hole surface recombination velocity at the emitter and p_{ec0} is the equilibrium hole concentration at the emitter contact.

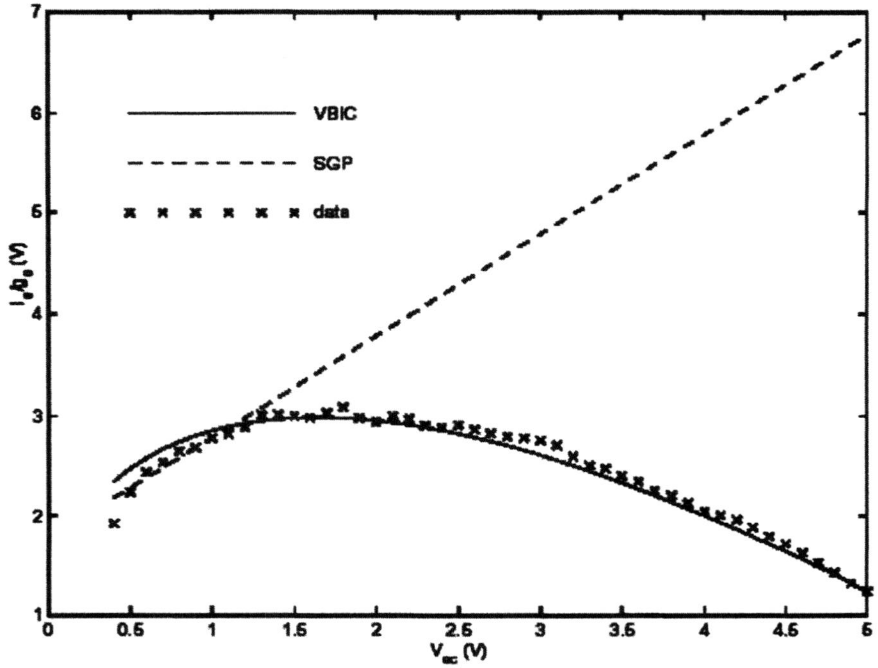

Figure 3. Early effect modeling of VBIC and SGP

For a shallow emitter, equating the surface recombination current density for holes to the hole diffusion current from the edge of the base gives

$$qS_h p_{ec} \approx qD_h \frac{\partial p}{\partial x} \approx qD_h \frac{n_{ie}^2 \exp(V_{bei}/V_{tv})/N_d - p_{ec}}{w_e} \tag{25}$$

where D_h is the hole diffusion constant, N_d is the doping density in the emitter (so the equilibrium hole concentration is nearly n^2_{ie}/N_d), and w_e is the depth of the emitter. This gives

$$p_{ec} \approx \frac{n_{ie}^2 \exp(V_{bei}/V_{tv})}{N_d(1 + S_h w_e/D_h)} \tag{26}$$

and thus the surface recombination current is close to proportional to $\exp(V_{bei}/V_{tv})$.

For recombination in the quasi neutral emitter $n \approx N_d \gg p$, and , $np \gg n^2_{ie}$, therefore

$$R_{srh} \approx \frac{p}{\tau_h},\ R_{aug} \approx c_e n^2 p \tag{27}$$

and because $p \sim \exp(V_{bei}/V_{tv})$ the quasi neutral region recombination current is also close to proportional to $\exp(V_{bei}/V_{tv})$.

In the base-emitter space charge region there is little Auger recombination (this process involves 3 interacting mobile carriers and is only likely in regions of high carrier concentrations), so Shockley-Read-Hall recombination dominates $\phi_e \approx 0$ and $\phi_h \approx V_{bei}$ in this region, so from equations (3), (4), and (22),

$$R_{srh} \approx \frac{n_{ie}\exp(V_{bei}/V_{tv})}{\tau_h(\exp(\psi/V_{tv})+1)+\tau_e(\exp((V_{bei}-\psi)/V_{tv})+1)}. \tag{28}$$

This rate is maximized for

$$\psi = \frac{V_{bei} - V_{tv}\log(\tau_h/\tau_e)}{2} \tag{29}$$

and for $\tau_h \approx \tau_e$ has a value

$$R_{srh} \approx \frac{n_{ie}\exp(V_{bei}/(2V_{tv}))}{\tau_h + \tau_e}. \tag{30}$$

The space charge recombination current is therefore close to proportional to $\exp(V_{bei}/(2V_{tv}))$.

Based on the above physical analyses, the base-emitter component of the intrinsic transistor base current in VBIC is modeled as

$$I_{be} = W_{BE}\left(I_{BEI}\left(\exp\left(\frac{V_{bei}}{N_{EI}V_{tv}}\right)-1\right)+I_{BEN}\left(\exp\left(\frac{V_{bei}}{N_{EN}V_{tv}}\right)-1\right)\right) \tag{31}$$

which includes both an ideal component, modeled with a saturation current I_{BEI} and ideality factor $N_{EI} \approx 1$, that comprises the emitter contact and quasi neutral region recombination, and a nonideal component for the space charge region component, modeled with saturation current I_{BEIN} and ideality factor $N_{EN} \approx 2$. The ideality factors are treated as model parameters, and can be quite different from the values of 1 or 2 for HBTs. The base-collector component is similarly modeled as

$$I_{bc} = I_{BCI}\left(\exp\left(\frac{V_{bci}}{N_{CI}V_{tv}}\right)-1\right)+I_{BCN}\left(\exp\left(\frac{V_{bci}}{N_{CN}V_{tv}}\right)-1\right). \tag{32}$$

The extrinsic base-emitter recombination current is

$$I_{bex} = W_{BE}\left(I_{BEI}\left(\exp\left(\frac{V_{bex}}{N_{EI}V_{tv}}\right)-1\right)+I_{BEN}\left(\exp\left(\frac{V_{bex}}{N_{EN}V_{tv}}\right)-1\right)\right). \tag{33}$$

The weak avalanche current I_{gc} is (Kloosterman, 1988)

$$I_{gc} = (I_{tzf}-I_{tzr}-I_{bc})A_{VC1}V_{gci}\exp(-A_{VC2}V_{gci}^{M_C-1}) \tag{34}$$

where A_{VC1} and A_{VC2} are model parameters, and V_{gci} is $P_C - V_{bci}$ limited, in C_∞ a continuous manner, to be greater than 0.

The intrinsic base resistance R_{BI} is modulated by the normalized base charge q_b. This accounts both for the base width variation from the Early effect, and the decrease in resistivity from increased mobile carrier concentration under high level injection conditions. VBIC does not include the I_{RB} emitter crowding modulation model of SGP. This effect can be taken into account, to a first order, by using the parameter W_{BE} to partition some of the base-emitter component of base current to I_{bex}, which is "external" to R_{BI}. This does not work well over all biases, however a simple model of emitter crowding, consistent for both DC and AC modeling, has not yet been developed.

If the model is biased so that the base region becomes depleted of charge, the modulated base resistance R_{BI}/q_b can become very large. q_b is limited to a lower value of 10^{-4} in VBIC (in a C_∞ continuous manner), but this is still sufficiently small to allow the model to support an unrealistically high V_{be} during a transient simulation for a device coming out of having a depleted base region. Multidimensional effects cause the device to turn on in a distributed manner from the edge of the emitter under such circumstances, and partitioning some of the baseemitter component of base current to I_{bex} prevents modeling the unrealistically high V_{be} values.

The parasitic transistor is modeled similarly to the intrinsic transistor.

$$I_{tfp} = I_{SP}\frac{W_{SP}\exp\left(\frac{V_{bep}}{N_{FP}V_{tv}}\right)+(1-W_{SP})\exp\left(\frac{V_{bci}}{N_{FP}V_{tv}}\right)-1}{q_{bp}} \tag{35}$$

$$I_{trp} = I_{SP}\frac{\exp(V_{bcp}/(N_{FP}V_{tv}))-1}{q_{bp}} \tag{36}$$

where the parasitic normalized base charge includes only a forward high level injection component,

$$q_{bp} = 1 + \frac{I_{SP}\left(W_{SP}\exp\left(\frac{V_{bep}}{N_{FP}V_{tv}}\right) + (1 - W_{SP})\exp\left(\frac{V_{bci}}{N_{FP}V_{tv}}\right) - 1\right)}{q_{bp}I_{KP}}. \tag{37}$$

where I_{SP}, N_{FP}, and I_{KP} are the saturation current,. ideality factor, and knee current for the parasitic transistor. The biases V_{bep} and V_{bcp} are between nodes bx and bp, and si and bp, respectively. The partitioning factor W_{SP} can be used to select a base-emitter control bias for the parasitic transistor either as shown in Figure 1, between nodes bx and bp, or from the base-collector of the intrinsic transistor, between nodes bi and ci. The structure of particular transistor determines which is more appropriate.

Although VBIC does not include a complete Gummel-Poon transistor for the parasitic, it does model the most important aspects of the behavior of this device. The transport current, including high level injection, models the substrate current when the intrinsic transistor goes into saturation. This is not included in the SGP model, yet is critical for accurate modeling of BJT behavior in saturation. The parasitic base-collector charge Q_{bcp} is important for modeling collector-substrate capacitance. And although it normally should be reverse biased, the parasitic base-collector base current component I_{bcp} is important for detecting any inadvertent forward biasing of the base-collector junction. The parasitic base-emitter components are nearly in parallel with the intrinsic base-collector components, although the former are still useful for accurate modeling of the distributed nature of devices. The parasitic transistor modeling is completed with the modulated parasitic base resistance R_{BIP} / q_{bp}, and the parasitic base-emitter and base-collector junction charges, both of which include depletion components, and the former also has a diffusions component modeled via the reverse transit time T_R of the intrinsic transistor.

$$i0 = \frac{V_{rci} + V_{tv}\left(K_{bci} - K_{bcx} - \log\left(\frac{K_{bci} + 1}{K_{bcx} + 1}\right)\right)}{R_{CI}} \tag{38}$$

$$ci = \sqrt{1 + G_{AMM}\exp\left(\frac{V_{bci}}{V_{tv}}\right)},\ K_{bcx} = \sqrt{1 + G_{AMM}\exp\left(\frac{V_{bcx}}{V_{tv}}\right)} \tag{39}$$

One of the major deficiencies of the SGP model is its lack of modeling of quasisaturation, when the base pushes into, and modulates the conductivity of, the collector. The Kull model (Kull, 1985) is the most widely accepted basis for quasisaturation modeling. However, this model can exhibit a negative output conductance at high V_{be}, see Figure 4. VBIC modifies the Kull model to avoid the negative output conductance problem, and includes an empirical model of the increase of collector current at high bias. The Kull quasi-saturation model without velocity saturation is

$$I_{epi0} = \frac{I_{epi0}}{\sqrt{1 + \left(\frac{R_{CI} I_{epi0}/V_O}{1 + \sqrt{0.01 + V_{rci}^2/(2 V_O H_{RCF})}}\right)^2}}. \tag{40}$$

where V_{bcx} is the extrinsic base-collector bias, between nodes bi and cx of Figure 1, and $V_{rci} = V_{bci} - V_{bcx}$ is the bias across the intrinsic collector resistance R_{CI}.VBIC models the collector current as

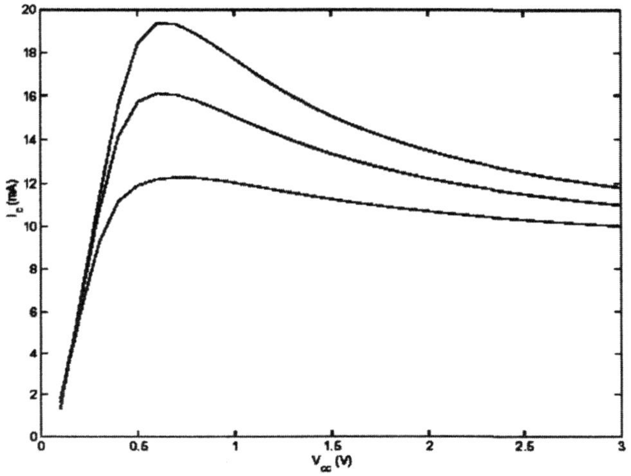

Figure 4. Negative output conductance from Kull model.

The temperature mappings of the VBIC parameters are as follows. All resistance temperature variations are modeled with the empirical mobility temperature relation (Jacoboni, 1977)

$$R(T) = R(T_{nom})\left(\frac{T}{T_{nom}}\right)^{X_R} \tag{41}$$

with separate exponents X_R for each of the emitter, base, collector, and substrate. The temperatures are in degrees Kelvin. The saturation currents vary with temperature as, for example for I_S,

$$I_S(T) = I_S(T_{nom})\left(\left(\frac{T}{T_{nom}}\right)^{X_{IS}} \exp\left(-E_A\frac{1 - T/T_{nom}}{V_{tv}}\right)\right)^{1/N_F} \tag{42}$$

where there is a separate exponent X_{IS} and activation E_A energy for each saturation current. The built-in potential and zero bias junction capacitance parameters are modeled over temperature similarly to the SGP model, with a modification avoid the built-in potential going negative for high temperatures.

N_F, N_R and A_{VC1} are modeled as having a linear temperature dependence. The epi doping parameter G_{AMM} is modeled over temperature as in equation (42), and the epi drift saturation voltage V_O is modeled over temperature as in equation (41).

The electrothermal modeling in VBIC follows the formulation of Vogelsong (1989) and McAndrew (1992). All of the branch constituent relations detailed above are modified to include a dependence on the local temperature rise, the voltage at the node dt, as defined in the temperature mappings above. This greatly complicates the modeling equations, however the procedure for doing this is completely automated ion VBIC, and is done using symbolic algebra software. The power dissipation is I_{th}

$$\begin{aligned} I_{th} = {} & I_{be}V_{bei} + I_{bex}V_{bex} + I_{bc}V_{bci} + (I_{tzf} - I_{tzr})(V_{bei} - V_{bci}) + I_{bep}V_{bep} \\ & + I_{bcp}V_{bcp} + (I_{tfp} - I_{trp})(V_{bep} - V_{bcp}) + I_{rcx}V_{rcx} + I_{rci}V_{rci} \\ & + I_{rbx}V_{rbx} + I_{rbi}V_{rbi} + I_{re}V_{re} + I_{rbp}V_{rbp} + I_{rs}V_{rs} - I_{gc}V_{bci} \end{aligned} \tag{43}$$

which is the sum of the products of branch currents and voltages over all elements of the VBIC equivalent network that do not store energy.

4. PARAMETER EXTRACTION

Because of the similarity of some parts of VBIC to SGP, some parts of the parameter extraction strategy for VBIC are similar to those for SGP (Parker, 1995). However, the additional modeling features of VBIC require additional extraction algorithms, and because, unlike SGP, the DC and AC (capacitance) models are linked in VBIC through the Early effect model the extraction of the Early voltages requires the junction depletion capacitances to be modeled.

The first step in VBIC characterization (parameter determination) is therefore to extract the junction depletion capacitance parameters. This is easily done by using nonlinear least squares optimization to fit measured *C* (*V*) data for each of the baseemitter, base-collector, and collector substrate junctions. The base-collector capacitance is partitioned between C_{JC} and C_{JEP} based on the relative geometries of the intrinsic (under the emitter) and extrinsic portions of the base-collector junction.

From forward output data at low V_{be} bias and reverse output data at low V_{bc} bias the output conductances normalized by current, g^f_o/I_c and g^r_o/I_e, are calculated, and then the solution of

$$\begin{bmatrix} q_{bcf}-c_{bcf}/(g^f_o/I_c) & q_{bef} \\ q_{bcr} & q_{ber}-c_{ber}/(g^r_o/I_e) \end{bmatrix} \begin{bmatrix} 1/V_{EF} \\ 1/V_{ER} \end{bmatrix} = \begin{bmatrix} -1 \\ -1 \end{bmatrix} \quad (44)$$

gives the VBIC Early voltages (McAndrew 1996). In equation (44) $q_{bef}(V^f_{be}, P_E, M_E)$ and $q_{bcf}(V^f_{bc}, P_C, M_C)$ the normalized base-emitter and basecollector depletion charges for the forward bias case, respectively, $q_{ber}(V^r_{be}, P_E, M_E)$ and $q_{bcr}(V^r_{bc}, P_C, M_C)$ are these charges for the reverse bias case, $c_{bcf}= \partial q_{bcf}/\partial V^f_{bc}$ and $c_{ber}= \partial q_{ber}/\partial V^r_{be}$.

The saturation currents and ideality factors for the various transport and recombination/generation currents can be extracted in the usual manner from the slopes and intercepts of the variation of the logarithms of the currents as functions of the applied voltages. The data need to be filtered to exclude high level injection and resistive debiasing effects. This is easily done by analyzing the derivative of the log(*I*) versus *V* data and excluding points that do not lie within some reasonable fraction, 5 to 10%, of its maximum value. The values obtained are then refined by optimization to fit the low bias data, both ideal and nonideal components. The activation energies for all saturation currents are determined by optimizing the fit to measured data, again filtered to keep only low biases, taken over temperature.

The knee currents can be determined as the current level at which the current gain drops to half its value.

Existing methods can be used to obtain initial values for the resistances. This can be difficult, and it is desirable to include both DC and AC data. Many of the simple procedures that have been proposed for BJT resistance calculation are based on oversimplifications of the model, and do not give realistic values. Optimization is used to refine the initial values, again preferably using DC and AC data. The quasisaturation parameters are likewise obtained by optimization to output curves that show significant quasi-saturation effects. Other parameters. such as knee currents and Early voltages, should also be refined in this optimization.

The avalanche model parameters are optimized to fit the output conductance of data that is affected by avalanche.

Because VBIC has the same transit time model as SGP, the existing techniques for SGP transit time characterization are directly applicable to VBIC. However, the quasi-saturation model also affects high frequency modeling, via Q_{CO}, particularly where f_T falls rapidly with increasing I_c, so optimization is again used to fit the AC data.

Several techniques are available for characterizing the thermal resistance and capacitance. Physical calculation from layout can be used. However for R_{TH} if the electrical parameters are characterized at low bias and using pulse measurements (Dunn, 1996) then R_{TH} can be determined by optimizing the fit to high current data that shows significant self heating.

5. RELATIONSHIP BETWEEN SGP AND VBIC PARAMETERS

Although VBIC offers many advantages over SGP, it was intended to default to being as close to SGP as possible. The Early effect formulation is the principle difference in formulation of the two models, the other features of VBIC are additions that, with the default parameters, are not active. Therefore, the easiest way to get started with VBIC is to use SGP as a base, and then incrementally include the features that are of greatest benefit for a given application. To help this Table 2 lists simple mappings from SGP parameters to VBIC parameters.

The Early voltages are the only parameters for which there is no direct mapping from SGP to VBIC. Because the Early effect models differ, the bias dependence of output conductance g_o cannot be matched between VBIC and SGP. Therefore the VBIC Early voltage parameters are derived from the SGP Early voltage parameters V_{AF} and V_{AR} by matching g^f_o/I_c and g^r_o/I_e between the two models at specific values of forward bias, V^f_{be} and V^f_{bc}, and reverse bias, V^r_{bc} and V^r_{be} (McAndrew, 1996). From the SGP model

Table 2. Mappings from SGP to VBIC parameters

VBIC	Mapping	VBIC	Mapping	VBIC	Mapping
R_{CX}	R_C	M_C	M_{JC}	X_{TF}	X_{TF}
R_{CI}	0	C_{JCP}	C_{JS}	V_{TF}	V_{TF}
R_{BX}	R_{BM}	P_S	V_{JS}	I_{TF}	I_{TF}
R_{BI}	$R_B - R_{BM}$	M_S	M_{JS}	T_R	T_R
R_E	R_E	I_{BEI}	I_S/B_F	T_D	$\pi\, T_F P_{TF}/180$
I_S	I_S	N_{EI}	N_F	E_A	E_G
N_F	N_F	I_{BEN}	I_{SE}	E_{AIE}	E_G
N_R	N_R	N_{EN}	N_E	E_{AIC}	E_G
F_C	F_C	I_{BCI}	I_S/B_R	E_{ANE}	E_G
C_{JE}	C_{JE}	N_{CI}	N_R	E_{ANC}	E_G
P_E	V_{JE}	I_{BCN}	I_{SC}	X_{IS}	X_{TI}
M_E	M_{JE}	N_{CN}	N_C	X_{II}	$X_{TI} - X_{TB}$
C_{JC}	$R_{JC}X_{CJC}$	I_{KF}	I_{KF}	X_{IN}	$X_{TI} - X_{TB}$
C_{JEP}	$C_{JC}(1 - X_{CJC})$	I_{KR}	I_{KR}	K_{FN}	K_F
P_C	V_{JC}	T_F	T_F	A_{FN}	A_F

$$\frac{g_o^f}{I_c} = \frac{1/V_{AF}}{1 - V_{be}^f/V_{AR} - V_{bc}^f/V_{AF}} \tag{45}$$

$$\frac{g_o^r}{I_e} = \frac{1/V_{AR}}{1 - V_{be}^r/V_{AR} - V_{bc}^r/V_{AF}} \tag{46}$$

are calculated, and then equation (44) is solved for V_{EF} and V_{ER}.

There is one other difference between the default parameters for VBIC and SGP. The F_C parameter, that limits how close to the built-in potential the junction voltage can go, for depletion charge and capacitance calculation, is 0.5 for SGP. This is too low and does not allow reasonable modeling of depletion capacitance into moderate forward bias. The VBIC default value is 0.9.

6. VBIC DC MODELING

Figures 5 through 7 compare DC modeling of VBIC to SGP. The improved accuracy of modeling the quasi-saturation region is apparent, as is the improved modeling of output conductance. The Early effect model, modulated collector resistance model, and weak avalanche model all contribute to the improvement in VBIC compared to SGP.

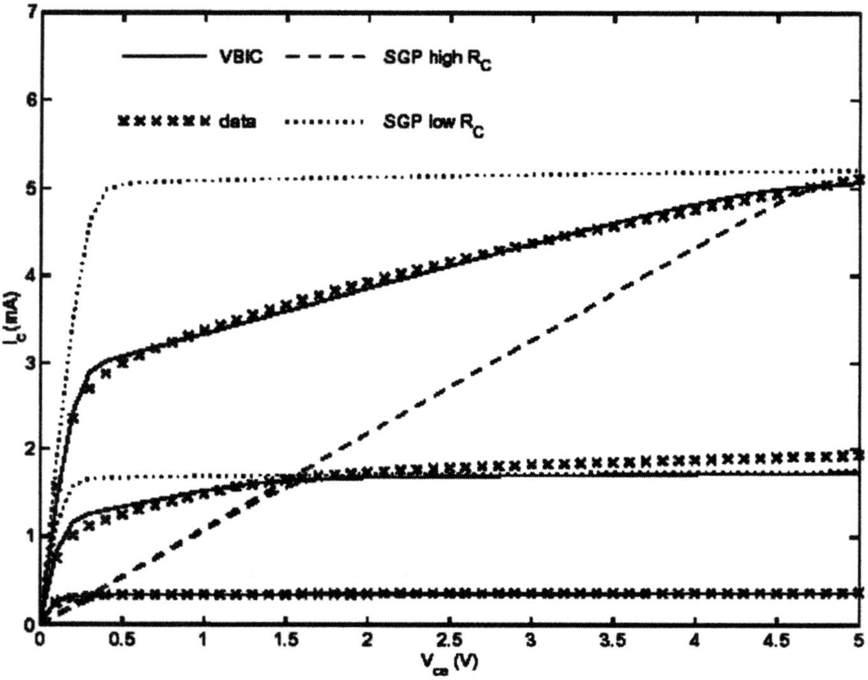

Figure 5. Forward output data with significant quasi-saturation.

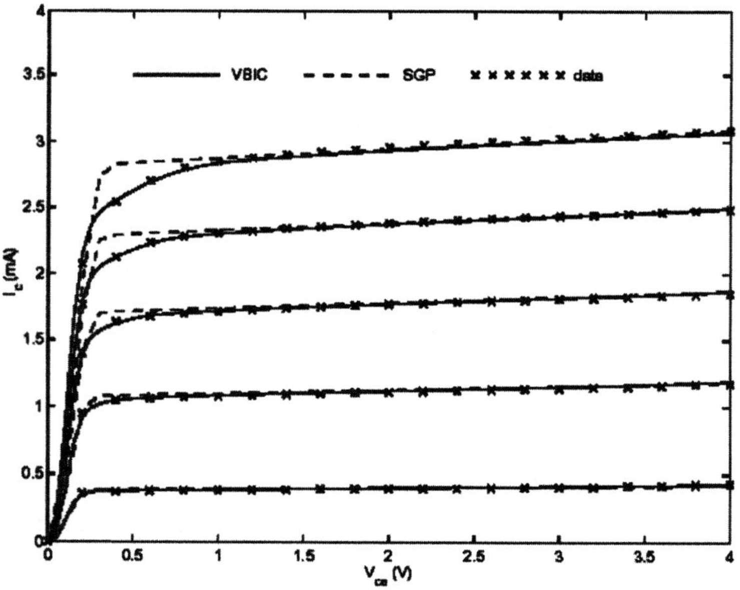

Figure 6. Forward output modeling comparison.

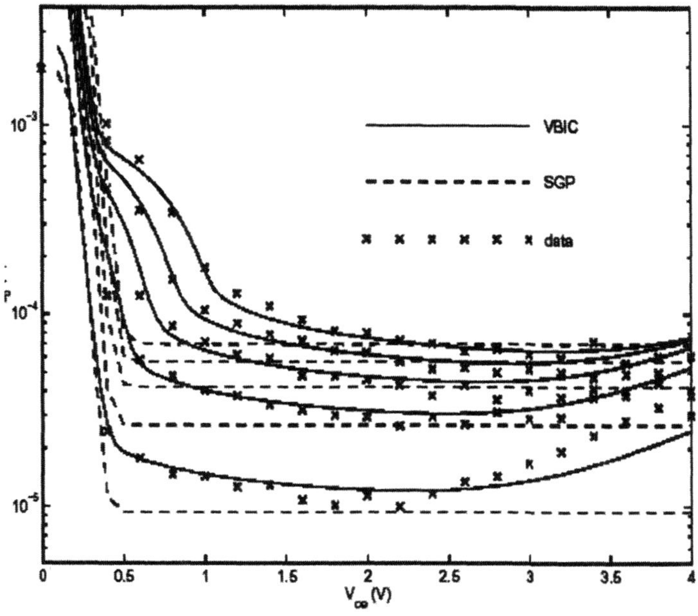

Figure 7. Output conductance modeling, from data of figure 6

7. ELECTROTHERMAL EXAMPLES

The coupled electrothermal (self heating) modeling capability is a major feature of VBIC. Self heating is significant for GaAs HBTs, because the thermal conductivity of GaAs is relatively low. In contrast to silicon BJTs, GaAs HBTs have a negative temperature coefficient for $\beta = I_c / I_b$. This means that for a fixed base current drive the collector current I_c decreases as temperature increases. Figure 8 shows output characteristics of a GaAs HBT with I_b is swept from 20 to 100μ**A** in steps of 20μA. At low I_c there is little self heating, but at high I_c there is significant self heating, which causes β and therefore I_c to decrease, which causes the output conductance g_o to become negative.

Self heating can also cause output resistance R_o degradation in silicon BJTs. Figure 9 shows this degradation, and for amplifiers operated at high current densities this can cause the small signal gain to decrease by a factor of up to about 3.

8. HIGH FREQUENCY MODELING

Although VBIC maintains, for the present, the forward transit time model of SGP, it still has improved high frequency modeling because of the improvements in other parts of the model. Figures 10 through 17 through show fits of VBIC to measured *s*-parameter data, for the listed values of V_{be} and V_{ce} varying from 1.0 to 3.0V in steps of 0.5V. The accuracy of VBIC is clear. Table 3 compares the RMS errors, over bias and frequency, in fitting the *s*-parameter data, between VBIC and SGP. The models were optimized in the same optimization tool, using the same optimization strategy, to the same data. The improvement in fit is again apparent.

Table 3. Comparison of SGP and VBIC fits to s-parameter data

Parameter	SGP RMS Re error (%)	VBIC RMS Re error (%)	SGP RMS Im error (%)	VBIC RMS Im error (%)
s_{11}	99.9	7.0	65.2	6.2
s_{12}	112.6	12.0	34.6	6.9
s_{21}	209.6	14.3	81.6	8.3
s_{22}	31.3	8.5	63.8	8.5

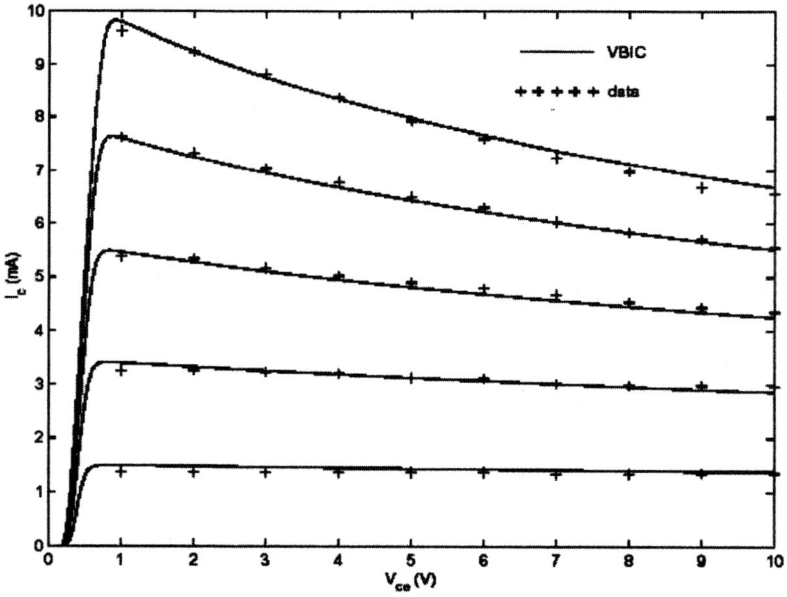

Figure 8. GaAs HBT electrothermal modeling with VBIC

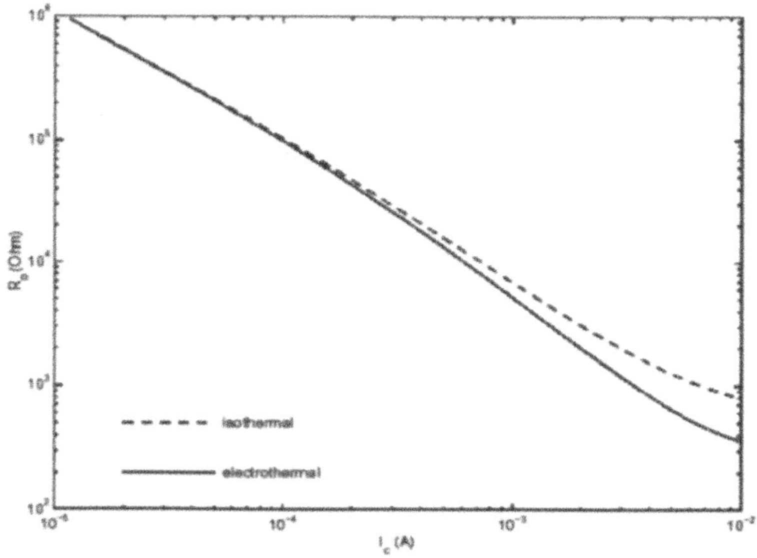

Figure 9. Output resistance degration caused by self heating

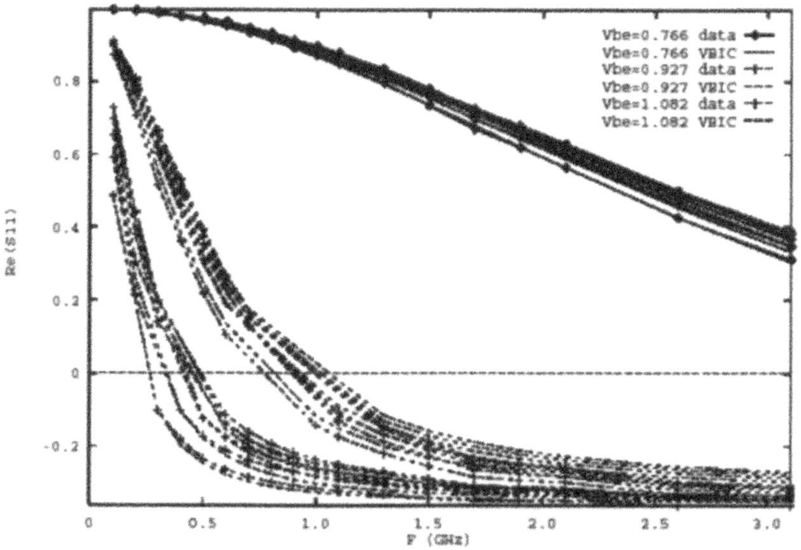

Figure 10. VBIC modeling of the real part of S_{11}.

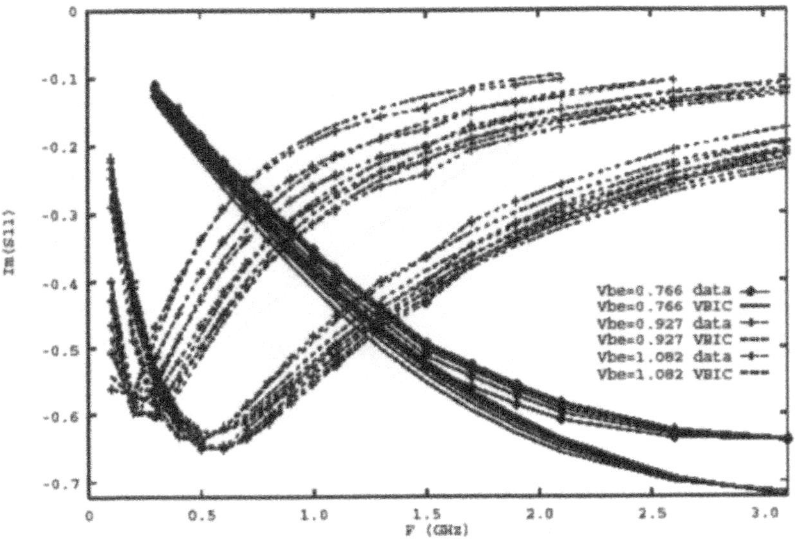

Figure 11. VBIC modeling of the imaginary part of S_{11}.

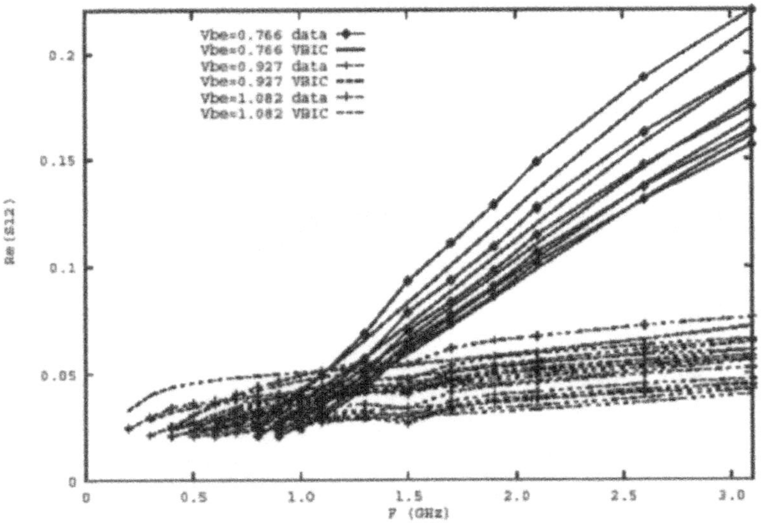

Figure12. VBIC modeling of the real part of S_{12}.

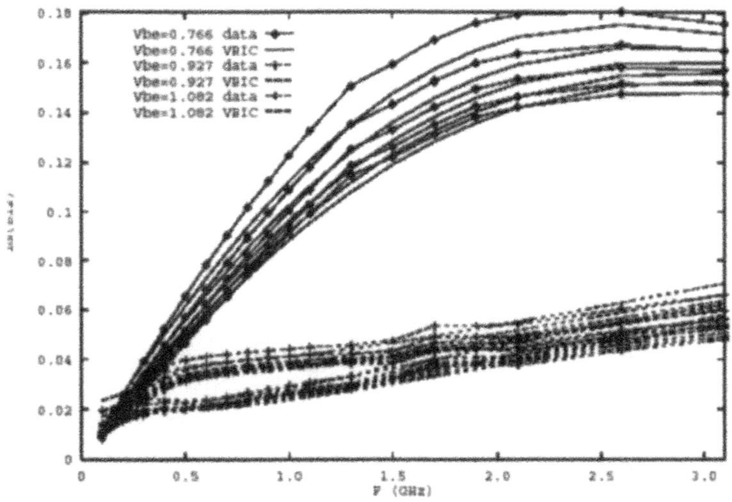

Figure 13. VBIC modeling of the imaginary part of S_{12}.

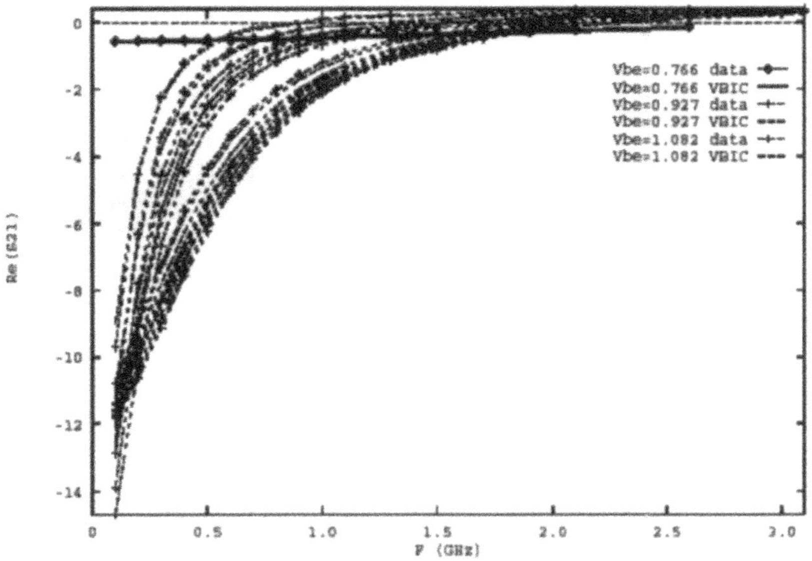

Figure 14. VBIC modeling of the real part of S_{21}.

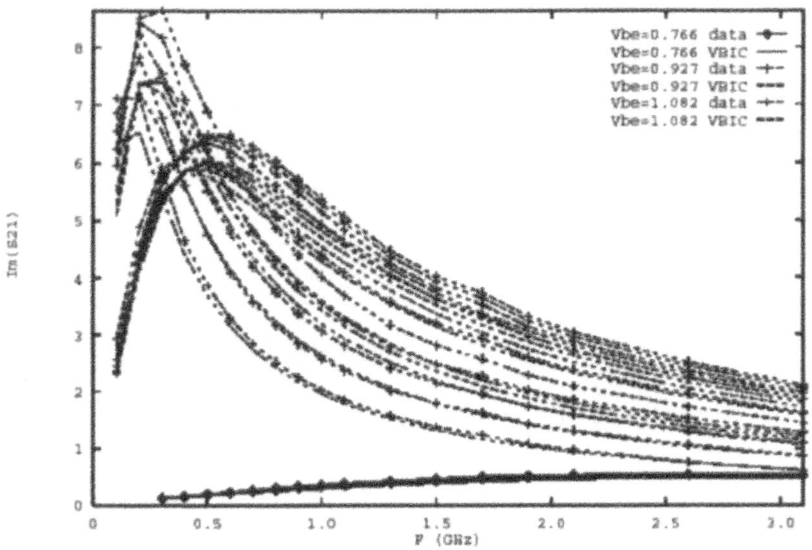

Figure15. VBIC modleing of the imaginary part of S_{21}.

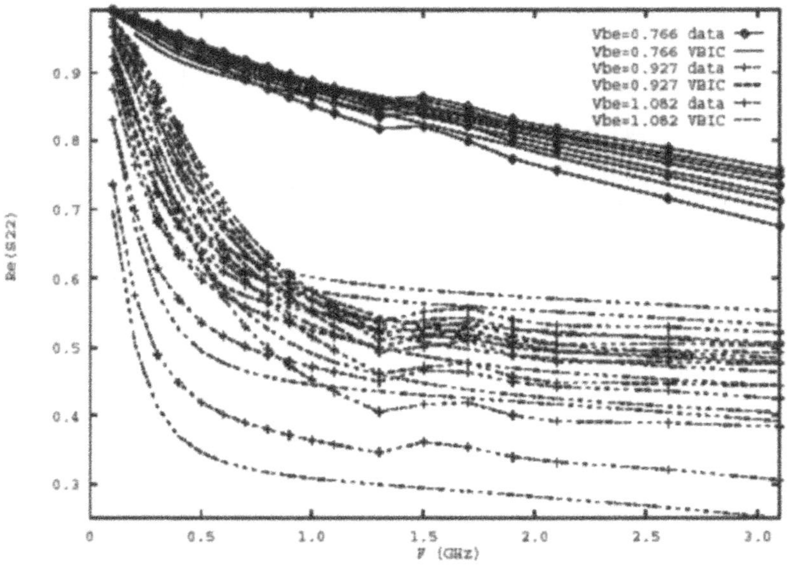

Figure 16. VBIC modeling of the real part of S_{22}

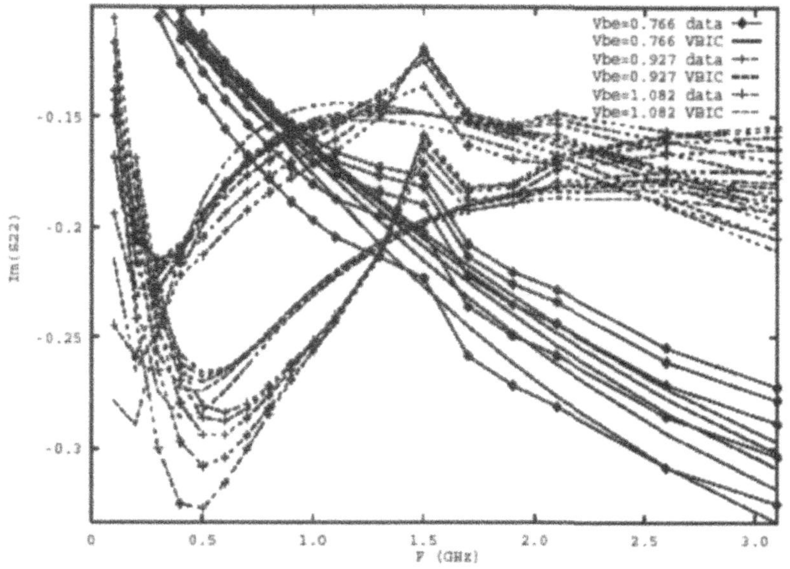

Figure 17. VBIC modeling of the imaginary part of S_{22}

9. CONCLUSIONS

This tutorial has reviewed the VBIC model, and provided details of the equations used within VBIC. Examples of improved modeling aspects of VBIC have been given. VBIC is at present undergoing some minor changes, and the only major change that is at present being considered is an improved transit time. A VBIC distribution package is available electronically at

`http://www-sm.rz.fht-esslingen.de/institute/iafgp/neu/VBIC/index.html`

and this includes complete source code, a pseudo-code description of the model, test solvers, a program that maps SGP to VBIC model parameters, and other information related to VBIC.

10. ACKNOWLEDGMENTS

Many people have contributed to, and are continuing to contribute to, VBIC, including Jerry Seitchik, Derek Bowers, Mark Dunn, Mark Foisy, Ian Getreu, Marc McSwain, Shahriar Moinian, Kevin Negus, James Parker, David Roulston, Michael Schröter, Shaun Simpkins, Paul van Wijnen, and Larry Wagner.

11. REFERENCES

de Graaff, H. C. and Kloosterman, W. J. (1985) New formulation of the current and charge relations in bipolar transistors for modeling for CACD purposes. *IEEE Trans. ED*, **32**, 2415-9.

Getreu, I. E. (1976) Modeling the Bipolar Transistor. Tektronix, Beaverton. Gummel, H. K., and Poon, H. C. (1970) An integral charge control model of bipolar transistors. *Bell Syst. Tech. J.*, **49**, 827-52.

Jacoboni, C., Canali, C., Ottaviani, G., and Alberigi Quaranta, A. (1977) A review of some charge transport properties of silicon. *Solid-State Electron.*, **20**, 77-89.

Jeong, H. and Fossum, J. G. (1989) A charge-based large-signal bipolar transistor model for device and circuit simulation. *IEEE Trans. ED*, **36**, 124-31.

Kloosterman, W. J. and de Graaff, H. C. (1988) Avalanche multiplication in a compact bipolar transistor model for circuit simulation. *Proc. IEEE BCTM*, 103-6.

Kull, G. M., Nagel, L. W., Lee, S.-W., Lloyd, P., Prendergast, E. J., and Dirks, H. K. (1985) A unified circuit model for bipolar transistors including quasi-saturation effects. *IEEE Trans. ED*, **32**, 1103-13.

McAndrew, C. C. (1992) A complete and consistent electrical/thermal HBT model. *Proc. IEEE BCTM*, 200-3.

McAndrew, C. C. and Nagel, L. W. (1996) Early effect modeling in SPICE. *IEEE JSSC*, **31**, 136-8.

Nagel, L. W. (1975) SPICE2: A computer program to simulate semiconductor circuits. Memo. no. ERL-520, Electronics Research Laboratory, University of California, Berkeley.

Parker, J. and Dunn, M. (1995) VBIC95 bipolar transistor model and associated parameter extraction. *HP EESof IC-CAP User's Meeting.*

Schaefer, B. and Dunn, M. (1996) Pulsed measurements and modeling for electro-thermal effect. *Proc. IEEE BCTM*, 110-7.

Stubing, H. and Rein, H.-M. (1987) A compact physical large-signal model for highspeed bipolar transistors at high current densities-Part I: one-dimensional model. *IEEE Trans. ED*, **34**, 1741-51.

Turgeon, L. J. and Mathews, J. R. (1980) A bipolar transistor model of quasisaturation for use in computer-aided design (CAD). *Proc. IEEE IEDM*, 394-7.

Vogelsong, R. S. and Brzezinski, C. (1989) Simulation of thermal effects in electrical systems. *Proc. IEEE APEC*, 353-6.

Weil, P. B. and McNamee, L. P. (1978) Simulation of excess phase in bipolar transistors. *IEEE Trans. Circuits Syst.*, **2**, 114-6.

12. BIOGRAPHY

Colin McAndrew received the Ph.D. and M.A.Sc. degrees in systems design engineering from the University of Waterloo, Waterloo, Ont., Canada, in 1984 and 1982, and the B.E. degree in electrical engineering from Monash University, Melbourne, Vic., Australia, in 1978. Since 1995 he has been the Manager of the Statistical Modeling and Characterization Laboratory at Motorola, Tempe AZ. From 1987 to 1995 he was a Member of Technical Staff at AT&T Bell Laboratories, Allentown PA. From 1984 to 1987 and 1978 to 1980 he was an engineer at the Herman Research Laboratory of the State Electricity Commission of Victoria.

Chapter 3

A MOS Transistor Model for Mixed Analog-digital Circuit Design and Simulation

Matthias Bucher [1], Christophe Lallement[2], François Krummenacher[1], Christian Enz[3]

[1] *Swiss Federal Institute of Technology, Lausanne (EPFL), Electronics Laboratory, ELB-Ecublens, CH-1015 Lausanne, Switzerland, Phone: +41 21 693 3975, Fax: +41 21 693 36 40, Email:mathias.bucher matthias.bucherepflch*

[2] *ERM-PHASE/ENSPS, Parc d'Innovation, Bld. Sébastien Brant, F-67400 Illkirch, France, Phone: +33 388 63 95 07 Fax: +33 388 63 95 09, Email: lallem@erm1.u-strasbg.fr*

[3] *Swiss Center for Electronics and Microtechnology (CSEM), Jaquet Droz 1, CH-2007 Neuchâtel, Switzerland, Phone: +41 32 720 52 18, Fax: +41 32 720 57 42, Email: christian.enz@csem.ch*

Abstract -- In the design cycle of complex integrated circuits, the compact device simulation models are the privileged vehicle of information between the foundry and the designer. Effective circuit design, particularly in the context of analog and mixed analog-digital circuits using silicon CMOS technology, requires a MOS transistor (MOST) circuit simulation model well adapted both to the technology and to the designer's needs. The MOST model itself should also help portable design, since design-reuse becomes a major advantage in the fast development of new products. Clearly, the MOST model must be based on sound physical concepts, and be parameterized in such a way that it allows easy adaptation to very different CMOS technologies, and provides the designer with information on important parameters for design. This chapter describes an analytical, scalable compact MOST model, called 'EKV' MOST model, which is built on fundamental physical properties of the MOS transistor. Among the original concepts used in this model are the normalization of the channel current, and taking the substrate as a reference instead of the source. The basic long-channel model is formulated in symmetric terms of the source-to-bulk and drain-to-bulk voltages. In particular, the transconductance-to-current ratio is accurately described for all levels of current from weak inversion through moderate and to strong inversion. This characteristic is almost invariant with respect to process parameters and technology scaling; therefore, the model can be adjusted to a large range of different technologies. Short-channel effects have been included in the model for the simulation of deep submicron technologies. A full charge-based dynamic model as well as the thermal noise model are derived within the same approach. The continuity of the model characteristics is based on the use of a single equation, enhancing circuit convergence. The relative simplicity of the model and its low number of parameters also ease the process of parameter extraction, for which an original method is proposed. This MOST model is used in the context of low-voltage, low-current analog and analog-digital circuit design using deep submicron technologies. A version of this model based on the same fundamental concepts, is also available as a public-domain model in various commercially available simulators.

1. INTRODUCTION

The continuing decrease of the supply voltage to reduce the power consumption of digital circuits strongly affects the design of the analog part of a mixed analog/digital IC. As a consequence of the supply voltage reduction, the operating points of many MOS transistors forming the analog circuits move towards the region of moderate inversion. Low-voltage design of CMOS circuits, under supply voltages as low as 1V or below, typically requires operation in moderate inversion. During the design process, the operating points are very often set in terms of available drain current and targeted transconductances. Once the current and transconductances are chosen, the transconductance-to-current ratio is defined and the corresponding operating point can be fixed in terms of aspect ratio (or inversion coefficient [1]). Since the circuit performances directly depend on the devices' transconductances, a good control and prediction of the transconductances and the corresponding operating points is crucial, even if they fall in the moderate inversion region. This can easily be done by using the transconductance-to-current ratio modeling approach described in this chapter.

Most of the MOS transistor (MOST) models [2] currently available in the public domain have been developed starting from the large-signal strong inversion operation and then extended to weak inversion by different means which not always ensure the continuity and/or the accuracy of the characteristic in the moderate inversion. The transconductances may then be wrongly estimated which can result in serious design problems. The g_{ms} / I_D approach, which is the basis of the EKV MOST model, starts from the small-signal model, from which the large-signal static, dynamic as well as thermal noise models are derived by integration. This intrinsically ensures the continuity (including higher-order derivatives) of all the model characteristics.

The basic model features used in this approach and contrasting from those used in many other MOST models, can be summarized as follows:

- *bulk reference* for all voltages instead of source reference
- the *pinch-off voltage* V_P and *slope factor* n, are the principal model internal variables, and are both function mainly of the gate voltage
- use of *normalized current* as another model internal variable of central importance
- non-regional approach, using single equations for all operating regimes
- correct behavior in the asymptotic regions of *weak* and *strong inversion* operation
- correct prediction of all transconductances in *moderate inversion*
- symmetrical forward and reverse operation
- low number of parameters
- *hierarchical model structure* allowing to formulate simple hand calculation expressions.

The use of current normalization has numerous advantages. It is particularly useful for ratio-based design techniques as described in [3]. The normalized drain current is also the principal model variable used in this modeling approach.

The present contribution shows this approach to be valid over many generations of MOS technologies. New features extending the model to application with deep submicron technologies are presented. The normalized transconductance to current rario characteristic, which represents the foundation of the model, is presented in Section 2, and is used to derive the ideal long-channel static model. The latter allows simple hand-calculation expressions to be formulated [1], which are useful in general design practice [3] as well as in educational contexts [4]. Vertical field dependent mobility and non-uniform channel doping effects are included in the long-channel model. In Section 3, effects related to short and narrow device geometries are introduced. In particular, simple models for the reverse short-channel effect (RSCE) and the bias dependent series resistances are presented. In Section 4 a dynamic model for the node charges as well as a thermal noise model are developed, using the normalized current as main variable. Some aspects of the model for computer simulation are discussed in Section 5. A complete parameter extraction method from DC measurements is presented, demonstrating the scalability of the model for submicron CMOS technologies.

2. THE LONG-CHANNEL MODEL

2.1 Transconductance-to-current ratio

Unlike many other MOST modeling approaches, the voltages are all referred to the local substrate instead of the source, to preserve the structural symmetry of the MOS device also in the model. Two transconductances can be defined, namely the gate transconductance,

$$g_{mg} \equiv \frac{\partial I_D}{\partial V_G} \tag{1}$$

and the transconductance from the source,

$$g_{ms} \equiv -\frac{\partial I_D}{\partial V_S} \tag{2}$$

The following relationship holds between the transconductances in saturation [1],

$$g_{ms} = n \cdot g_{mg} \tag{3}$$

The *slope factor* n is defined as [1]

$$n \equiv \left[\frac{\partial V_P}{\partial V_G}\right]^{-1} = 1 + \frac{\gamma}{2 \cdot \sqrt{\psi_0 + V_P}} \tag{4}$$

where the parameter $\psi_0 \equiv 2\phi_F + V_{ch}$ is the approximation of the surface potential in strong inversion; ϕ_F is the Fermi potential of the majority carriers, $U_T = k \cdot T / q$ is the thermodynamic potential with k being the Boltzmann constant and q the unit charge. The parameter γ is the body or substrate effect factor

$$\gamma = \frac{\sqrt{2 \cdot q \cdot \varepsilon_{si} \cdot N_{sub}}}{C_{ox}'} \tag{5}$$

where N_{sub} is the substrate doping concentration in the channel; the gate oxide capacitance per unit area C_{ox} depends on the oxide thickness t_{ox} as

$C_{ox} = \varepsilon_{ox}/t_{ox}$, where ε_{si} and ε_{ox} denote the permittivity of silicon and silicon dioxide, respectively.

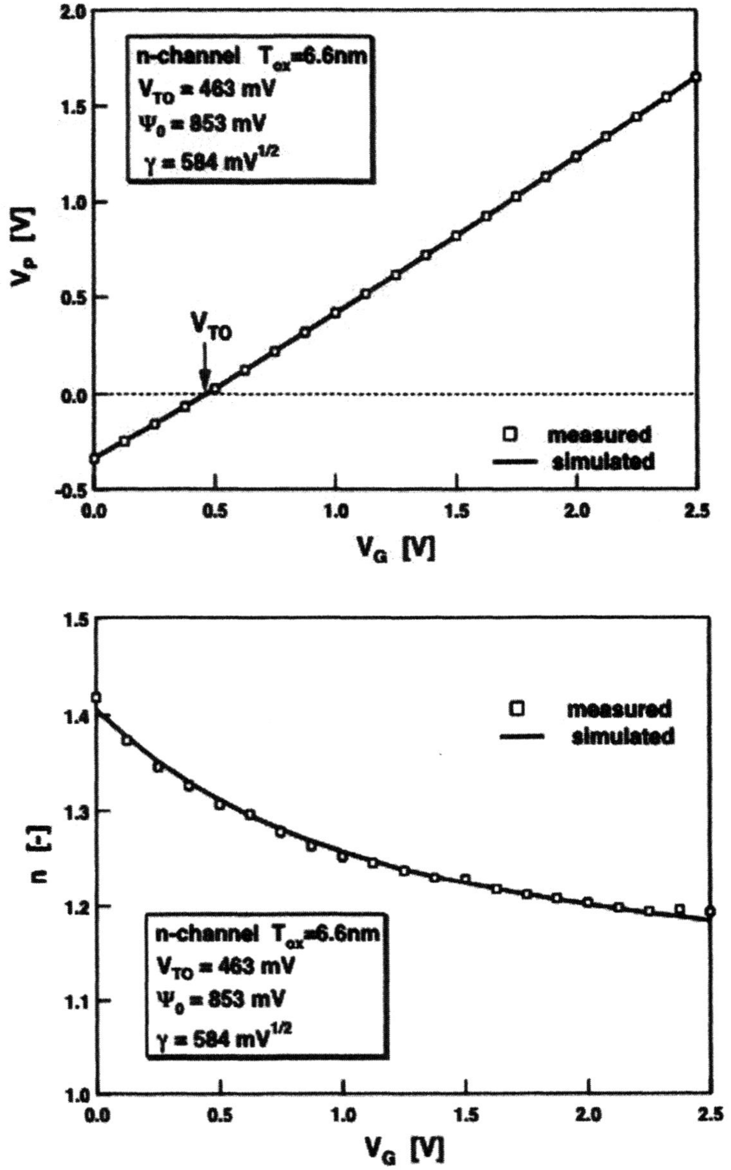

Figure1. Pinch-off voltage V_P (left) and slope factor n (right) vs. V_G measured (markers) on a long n-channel device of a 0.25 μm CMOS technology and simulated (lines). The threshold voltage V_{TO} corresponds to the intersection point $V_P = 0V$.

The *pinch-off voltage* V_P represents the channel voltage at a given gate voltage V_G, for which the inversion charge density Q_{inv} of the mobile charge forming the channel becomes negligible with respect to the depletion charge density Q_B [1]

$$V_P = V_G - V_{TO} - \gamma \cdot \left[\sqrt{V_G - V_{TO} + \left(\sqrt{\psi_0} + \frac{\gamma}{2} \right)^2} - \left(\sqrt{\psi_0} + \frac{\gamma}{2} \right) \right]. \tag{6}$$

The pinch-off voltage accounts for *threshold voltage* and *substrate effects* through the use of the parameters V_{TO} and γ, respectively. The parameters V_{TO} and ψ_0 are temperature dependent. An approximate expression for V_P which is useful for hand calculation [1],

$$V_P \cong \frac{V_G - V_{TO}}{n(V_G)} \tag{7}$$

shows that the substrate effect is accounted for through a function of the gate voltage V_G, unlike conventional MOST models where it is commonly expressed as a function of V_S. The pinch-off voltage (6) and the slope factor (4) are shown in Fig.1 versus the gate voltage V_G; measured characteristics are from a long n-channel device of a 0.25 μm CMOS technology.

The asymptotes of the g_{ms}/I_D characteristic can be found from the expressions of the transconductances in weak inversion [1]:

$$g_{mg} = \frac{I_D}{n \cdot U_T} \qquad g_{ms} = \frac{I_D}{U_T} \tag{8}$$

and in strong inversion

$$g_{mg} = \sqrt{\frac{2 \cdot \beta \cdot I_D}{n}} \qquad g_{ms} = \sqrt{2 \cdot n \cdot \beta \cdot I_D} \tag{9}$$

The normalized transconductance to current ratio in saturation can then be expressed through a single function valid in the entire region of inversion, the asymptotes of which are given by (8) and (9),

$$\frac{g_{mg} \cdot n \cdot U_T}{I_D} = \frac{g_{ms} \cdot U_T}{I_D} = G(i_f) = \begin{cases} 1 & (weak \quad inversion) \\ \dfrac{1}{\sqrt{i_f}} & (strong \quad inversion) \end{cases} \tag{10}$$

where i_f is the forward normalized drain current [1]. The transconductance to current ratio is plotted in Fig.2 versus the normalized current $i_f = I_F / I_S$ in saturation where $I_F \cong I_D$. The normalization factor I_S is called *specific current*. It is defined as the drain current corresponding to the intersection of the two asymptotes given by (10). I_S is related to the transistor aspect ratio [1]

$$I_S = 2 \cdot n \cdot \beta \cdot U_T^2 \tag{11}$$

where $\beta = \mu_{eff} \cdot C_{ox} \cdot W_{eff} / L_{eff}$ is the gain factor depending on the carrier mobility μ_{eff}. The specific current is a quantity allowing to delimit the regions of weak and strong inversion: $I_D << I_S$ (or $i_f << 1$) corresponds to weak inversion and $I_D >> I_S$ (or $i_f >> 1$) corresponds to strong inversion.

The normalized g_{ms}/I_D characteristic can be computed by numerically solving the Poisson and Gauss equations for the surface potential under long-channel and uniform doping assumptions. The result is plotted in Fig.2 versus the normalized current i_f for $\gamma = 0.7\sqrt{V}$. The sensitivity of the g_{ms}/I_D characteristic to γ is found to be small. A suitable function is required to describe this characteristic; in the past, various analytical expressions of varying degree of accuracy have been used [5][1]. A physics based expression has been proposed in [6],

$$G(i_f) = \frac{1}{\sqrt{\frac{1}{4} + i_f} + \frac{1}{2}} \tag{12}$$

derived from the assumption of a linear relationship between the surface potential ψ_s and inversion charge density Q_{inv} [7]. This expression has the advantage of analytical simplicity.

As shown in Fig.2, (12) is in excellent agreement with the results obtained from the numerical computation. Therefore, the present model is in excellent agreement with the theory.

Many trade-offs must be made in realistic technology scaling. As channel lengths decrease, the vertical dimensions, oxide thickness and junction depths, are decreased, while substrate doping is increased to overcome the detrimental impact of short-channel effects. At the same time, higher channel doping concentrations also lead to reduced low-field mobility of the carriers. Low-voltage operation requires as low threshold voltages as

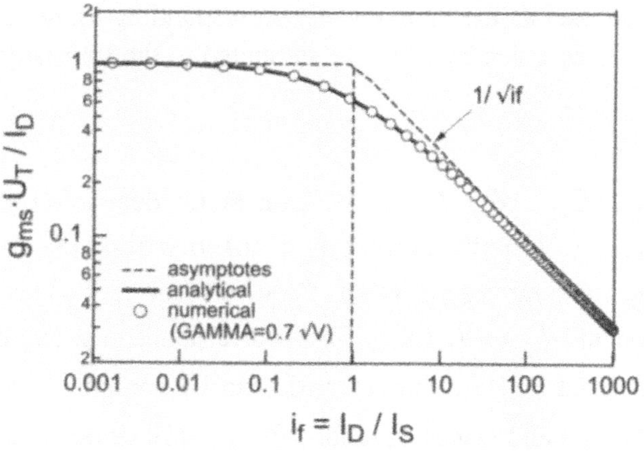

Figure 2. The normalized transconductance to current ratio $g_{ms} \cdot U_T / I_D$ versus normalized drain current I_D / I_S in saturation, for a long-channel device, computed by numerically solving the Poisson equation (markers) and analytical expression (12) (line). The intersection of the weak and strong inversion asymptotes defines the specific current I_S.

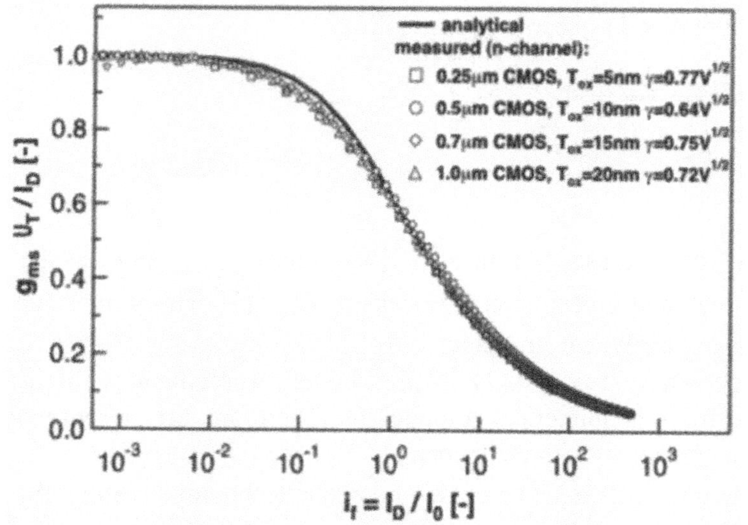

Figure 3. Normalized $g_{ms} \cdot U_T / I_D$ characteristics versus normalized current I_D / I_S in saturation, measured on long n-channel devices for three different standard CMOS technologies (markers) and compared to the analytical expression given by (12) (line).

possible, while high weak inversion slope as well as high threshold voltage are required to achieve low leakage currents. Minimum practical values for threshold voltage are in the range from $0.3V$ to $0.4V$, while supply voltages are reduced, to values of $1.5V$ and even $0.9V$ for future technologies, therefore reducing the available range of strong inversion operation. In any case, technology trends make operation in weak and moderate inversion become more and more important, and MOST models must give a precise description of these regions of operation. The present modeling approach based on transconductance to current ratio is suitable to address this situation.

Practical values for the important technological parameter γ lie in the range of $0.3\sqrt{V}$ to $0.9\sqrt{V}$. Noting that the substrate effect depends on oxide thickness and channel doping as $\gamma \sim t_{ox} \cdot \sqrt{N_{sub}}$, its dependence on technology scaling can be assessed, depending on the scaling rules adopted. Assuming constant field scaling (e.g. [8]) with $t_{ox} \sim \kappa^{-1}$ and $N_{sub} \sim \kappa$ when the scaling factor κ is increased, γ would decrease as $\gamma \sim \kappa^{-1/2}$ [8]. In realistic scaling, observed over several technology generations, assuming $\gamma \sim \kappa^{-0.77}$ and $N_{sub} \sim \kappa^{1.6}$ [9], the variation of γ is even smaller and it can be considered almost as constant. As long as the same physical effects are dominating, the normalized g_{ms}/I_D characteristic can be considered as independent of technology and scaling. In Fig.3, measurements of the normalized g_{ms}/I_D characteristic from three different CMOS technology generations and from three different foundries are shown. The measurements are made on long-channel devices, with minimum feature sizes ranging from $1\,\mu m$ to $0.5\,\mu m$. The normalized g_{ms}/I_D characteristic has been measured on many other technologies, confirming the excellent qualitative agreement with the analytical expression (12).

In the next subsections, the normalized g_{ms}/I_D characteristic is used to elaborate a complete model that is valid in all regions of operation. The large-signal static model, the charge-based dynamic model as well as the thermal noise model can all be derived from the same function (12), as will be shown in the following. Note that these following derivations have in their essence already been used to establish the former 'EKV' model formulations. In contrast to these, the present formulation is based on the expression (12), which has an improved physical basis and is analytically more tractable.

2.2 The static model for the drain current

The drain current I_D can be expressed as the difference between a forward current I_F and a reverse current I_R, depending only on $(V_P - V_S)$ and $(V_P - V_D)$, respectively [1]:

$$I_D = I_F - I_R = I_S \cdot (i_f - i_r) \tag{13}$$

In saturation (i.e. for $V_D >> V_P$), the reverse current becomes negligible compared to the forward current, and the drain current simply reduces to the forward current.

The source transconductance in saturation can be expressed as

$$g_{ms} \equiv -\left.\frac{\partial I_D}{\partial V_S}\right|_{V_G, V_D} = \frac{\partial I_F}{\partial (V_P - V_S)} = \frac{I_S}{U_T} \cdot \frac{\partial i_f}{\partial \nu} \tag{14}$$

where $\nu \equiv (V_P - V_{S(D)})/U_T$ is the normalized voltage for the forward (respectively reverse) current. From (12), the source transconductance in saturation is also given by

$$g_{ms} = \frac{I_F}{U_T} \cdot G(i_f) \tag{15}$$

Equating (14) and (15) and using (12) results in

$$\frac{\partial i_f}{\partial \nu} = i_f \cdot G(i_f) = \sqrt{\frac{1}{4} + i_f} - \frac{1}{2} \tag{16}$$

which can be integrated to find a large-signal interpolation function for the normalized voltage as a function of the normalized current

$$\nu = 2 \cdot \left(\sqrt{\frac{1}{4} + i} - \frac{1}{2}\right) + \ln\left(\sqrt{\frac{1}{4} + i} - \frac{1}{2}\right) \tag{17}$$

Eq. 17 needs to be inverted for the purpose of using the model for computer simulation, to obtain the normalized currents in terms of the normalized voltages. Unfortunately, (17) cannot be inverted analytically. A simple Newton-Raphson algorithm is used to obtain (18) [10][11]. It requires three iterations only for a maximum residual error below 10^{-6} when choosing adequate initial conditions. Little computational penalty is therefore incurred in using this simple numerical scheme. The inverted function will have the following asymptotes:

$$i = F(\nu) = \begin{cases} (\nu/2)^2 & (\nu >> 0) \\ e^{\nu} & (\nu << 0) \end{cases} \tag{18}$$

The normalized currents are now expressed as a function of the normalized voltages, for the whole current range from weak through moderate to strong inversion, as well as from conduction to saturation, as is required for the formulation of the compact MOST model.

2.3 Hand calculation model and circuit design

For analog circuit design, simplified model expressions can be obtained for the basic long-channel model. The simplified expressions for drain current and transconductances are obtained for the asymptotic regions of device operation, weak and strong inversion, conduction and saturation, and are summarized in Table I to Table III [1]. Apart from use of the hand calculation expressions in design, they often are also useful for the development of the parameter extraction method.

The relationship between the pinch-off voltage and the weak inversion slope should be noted, since it is important for model development and design practice. The long-channel expression for the drain current in weak inversion, using the approximate expression (7) for the pinch-off voltage and assuming $i_f >> i_r$ (requiring $V_D - V_S >\sim 5U_T$, in weak inversion), is given by

$$I_D \cong I_S \cdot \exp\left(\frac{V_P - V_S}{U_T}\right) \cong I_S \cdot \exp\left(\frac{V_G - V_{TO} - n \cdot V_S}{n \cdot U_T}\right) \tag{19}$$

The substrate effect thus determines the weak inversion slope through the substrate effect factor (or weak inversion slope factor) n. The subthreshold swing S_G, describing the change in gate voltage needed to change I_D by one decade, is also often used in this context, and is related to n simply as:

$$S_G \equiv \left[\frac{\partial(\log I_D)}{\partial V_G}\right]^{-1} = \ln(10) \cdot n \cdot U_T \cong 2.3 \cdot n \cdot U_T \tag{20}$$

An ideal slope factor of $n = 1$ therefore corresponds to the ideal subthreshold swing of $60mV/dec$. Similarly, the subthreshold swing for a modulation from the source can be defined as

$$S_S \equiv \left[\frac{\partial(\log I_D)}{\partial V_S}\right]^{-1} = \ln(10) \cdot U_T \cong 2.3 \cdot U_T \tag{21}$$

Note that no parameter specific to weak inversion has been introduced in the model so far. In general, the weak inversion slope is correctly predicted for long-channel transistors for modulation from both the gate and the source. This holds for technologies for which surface states are negligible, which is the case in most modern CMOS technologies.

Numerous applications of the model principles and its use in analog and mixed analog-digital circuit design can be found in [3]. In particular, a ratio-based design technique is described, which allows design of circuits that are insensitive to process variations and temperature to a first order. Using the specific current of a reference transistor, generated from a dedicated circuit, allows to set the inversion coefficient (or normalized current) of a device by simple scaling of this reference current by series and/or parallel combination of reference transistors. This ratio-based design technique is therefore attractive for designing circuits that can be ported from one technology to another, avoiding major redesign while preserving the main performance. The EKV model can therefore be considered as a tool enabling design-reuse and portability. Further aspects related to model formulation can also be found in [12].

The approach described so far offers the advantage of having a model with continuous drain current from weak to strong inversion and continuous n^{th}-order derivatives with respect to any terminal voltage. This is essential for the design and simulation of analog ICs and particularly for analog circuits that have to operate at a low supply voltage imposing that the operating points of many transistors be in the middle of the moderate inversion region. The continuity and accuracy of the derivatives is also important for the correct computation of distortion and intermodulation products which are fundamental limitations of RF circuits.

No mobility reduction effects, nor short-channel effects, have been taken into account so far. The long-channel model needs to be complemented to account for vertical field dependent mobility, for which an adequate expression will be developed in the next subsection. Another important aspect of precision MOS modeling is accounting for the effects of the non-uniformity of channel doping profiles with the depth from the Si-SiO_2 interface, resulting from ion implantation to reduce short-channel effects and to prevent punch-through. This effect is accounted for in the $V_P(V_G)$ relation as described in [13], requiring two additional parameters.

2.4 Vertical field dependent mobility

While oxide thickness is reduced when scaling CMOS technology, the power supply voltage is usually not reduced quite accordingly--as would be required by the constant field scaling--to maintain maximum speed for digital circuit applications. Thus higher fields result across the gate oxides in more advanced technologies [14].

Table 1: Drain current in strong inversion.

Mode	Expression for drain current	Condition
Conduction	$n \cdot \beta \cdot \left[V_P - \frac{V_S + V_D}{2} \right] \cdot (V_D - V_S)$ $\cong \beta \cdot \left[V_G - V_{TO} - \frac{n}{2} \cdot (V_S + V_D) \right] \cdot (V_D - V_S)$	$V_S \leq V_P$ $V_D \leq V_P$
Forward saturation	$\frac{n \cdot \beta}{2} \cdot (V_P - V_S)^2 \cong \frac{\beta}{2 \cdot n} \cdot (V_G - V_{TO} - n \cdot V_S)^2$	$V_S \leq V_P$ $V_D > V_P$
Reverse saturation	$-\frac{n \cdot \beta}{2} \cdot (V_P - V_D)^2 \cong \frac{-\beta}{2 \cdot n} \cdot (V_G - V_{TO} - n \cdot V_D)^2$	$V_S > V_P$ $V_D \leq V_P$
Blocked	$I_F = I_R \Rightarrow I_D = 0$	$V_D = V_S$

At high vertical fields, surface roughness scattering is considered as the main mechanism limiting mobility, while other scattering mechanisms dominate at lower fields. Mobility dependence on effective field has exposed a 'universal' behavior--regardless of doping levels--at high and intermediate vertical field strengths, when the effective mobility μ_{eff} is plotted with respect to the effective field E'_{eff}. The latter is expressed as a function of the depletion charge density Q_B and the inversion charge density Q_{inv} [15]:

$$E_{eff} = \left| \frac{Q_B' + \eta \cdot Q_{inv}'}{\varepsilon_{si}} \right| \tag{22}$$

Depending on the temperature range, the field strength and the type of carriers, mobility degradation can be expressed using different exponents of E'_{eff}. At high field strengths, a $\mu_{eff} \sim E'_{eff}$ [illegible] dependence is observed for holes (p-channel), while for electrons (n-channel) the dominant dependence is $\mu_{eff} \sim E'_{eff}$ [illegible].

Table 2: Drain current in weak inversion.

Mode	Expression for I_D	Condition
Conduction	$I_S \cdot e^{\frac{V_P}{U_T}} \cdot \left[e^{-\frac{V_S}{U_T}} - e^{-\frac{V_D}{U_T}} \right]$ $= I_S \cdot e^{\frac{V_P - V_S}{U_T}} \cdot \left[1 - e^{-\frac{V_D - V_S}{U_T}} \right]$ $\cong I_{D0} \cdot e^{\frac{V_G}{nU_T}} \cdot \left[e^{-\frac{V_S}{U_T}} - e^{-\frac{V_D}{U_T}} \right]$ $\cong I_{D0} \cdot e^{\frac{V_G - nV_S}{nU_T}} \cdot \left[1 - e^{-\frac{V_D - V_S}{U_T}} \right]$	$V_S > V_P$ $V_D > V_P$ $I_{D0} \equiv I_S \cdot e^{\frac{-V_{TO}}{n \cdot U_T}}$
Forward saturation	$I_F = I_S \cdot e^{\frac{V_P - V_S}{U_T}} \cong I_{D0} \cdot e^{\frac{V_G - n \cdot V_S}{n \cdot U_T}}$	$V_S > V_P$ $V_D > V_P$ $V_D - V_S >> U_T$
Reverse saturation	$-I_R = I_S \cdot e^{\frac{V_P - V_D}{U_T}} \cong I_{D0} \cdot e^{\frac{V_G - n \cdot V_D}{n \cdot U_T}}$	$V_S > V_P$ $V_D > V_P$ $V_S - V_D >> U_T$
Blocked	$I_F = I_R \Rightarrow I_D = 0$	$V_D = V_S$ or $V_S >> V_P$, $V_D >> V_P$

Accounting for these dependencies is particularly important for correct prediction of intermodulation. Also, a different factor for the dependency on inversion charge is observed, $\eta \cong 0.5$ for n-channel and $\eta \cong 0.3$ for p-channel [16] at room temperature. While most other MOST models use approximate and simplified expressions accounting for the global mobility dependence, the approach taken here is to consider the local effective field depending on the position, which is then integrated along the channel. The expression for the localized mobility dependence on vertical field, accounting for the two terms of mobility dependence on effective field as discussed above, is formulated as [11]

Table 3: Transconductances in strong and weak inversion.

	Strong inversion		Weak inversion
	Conduction	Forward saturation	
g_{mg}	$\beta \cdot (V_D - V_S)$	$\beta \cdot (V_P - V_S) = \sqrt{\frac{2 \cdot \beta \cdot I_D}{n}} = \frac{2 \cdot I_D}{n \cdot (V_P - V_S)} \cong \frac{2 \cdot I_D}{V_G - V_{TO} - n \cdot V_S}$	$\frac{I_D}{n \cdot U_T}$
g_{ms}	$n \cdot \beta \cdot (V_P - V_S) = \sqrt{2 \cdot n \cdot \beta \cdot I_F}$ $\frac{2 \cdot I_F}{V_P - V_S} \cong \frac{2 \cdot n \cdot I_F}{V_G - V_{TO} - n \cdot V_S}$		$\frac{I_F}{U_T}$
g_{md}	$n \cdot \beta \cdot (V_P - V_D) = \sqrt{2 \cdot n \cdot \beta \cdot I_R} = \frac{2 \cdot I_R}{V_P - V_D} \cong \frac{2 \cdot n \cdot I_R}{V_G - V_{TO} - n \cdot V_D}$	0	$\frac{I_R}{U_T}$

$$\mu_{eff}(x) = \frac{\mu_0}{1 + \frac{E_{eff}(x)}{E_1} + \left(\frac{E_{eff}(x)}{E_2}\right)^2} \tag{23}$$

where E_1 and E_2 are the model parameters related to the first- and second-order effective fields, respectively. The above expression (23) is comparable to the model e.g. as discussed in [17], which however does not consider local field dependent mobility. To account for mobility reduction globally throughout the whole channel, (23) is integrated along the channel [11], using the expressions for the charge densities in terms of the normalized current instead of position, Q_{inv} (i) and Q_B (i). The integration can be carried out, following the procedure outlined in Section 4 for the charges integration. The resulting mobility expression, in terms of the normalized currents i_f and i_r (which will be omitted here for brevity), will also depend on the longitudinal field, i.e. V_{DS} : mobility reduction due to vertical field is attenuated in saturation, since a part of the channel near the drain is less strongly inverted than at the source. Unlike other MOST models, the 2nd-order term of mobility reduction used here represents the fully integrated form of the local effective field dependence. This mobility model has shown very accurate results for technologies with thin gate oxides in strong inversion, both for n-channel transistors where the 2nd-order term is privileged, and for p-channel transistors adequately modeled by using the 1st-order term. A slight temperature dependence of the related parameters can be observed. This dependence can be neglected in a first-order approximation for usual temperature ranges from 0°C to 100°C, using only the temperature dependence of the low-field mobility .

Another point should be noted for this formulation of the mobility model. No specific parameters are introduced for the dependence on substrate bias, as it is often necessary in other models. The choice of a fixed parameter η has proven to simplify parameter extraction. Good results are obtained for the temperature range considered above--for both n- and p-channel--for different substrate biases, providing the substrate effect related parameters have been correctly extracted. Also, the asymptotic behavior of this model at very high fields is correct. Unphysical predictions, such as increasing drain current with increased substrate bias, as can sometimes be observed in other models, are avoided.

3. THE STATIC MODEL FOR SHORT AND NARROW GEOMETRIES

The previous derivation has been done assuming a long-channel device. In short-channel devices, the two-dimensional nature of the fields would actually also require a two-dimensional analysis. For the purposes of a circuit simulation model which should remain reasonably simple, approximations can be used, yielding a model with sufficient accuracy and efficiency.

For short-channel devices in strong inversion, the effects of velocity saturation and channel length modulation (CLM) are introduced. Another important effect affecting short-channel devices is the resistivity of the source and drain diffusion regions, which is typically bias dependent in technologies using lightly doped drain (LDD) structures.

In weak inversion, the short- and narrow-channel effects are introduced by modifying the expression of the pinch-off voltage V_P using a charge-sharing approach. The pinch-off voltage V_P for a long-channel device only depends on the gate voltage, whereas for short- and narrow-channel devices it becomes also a function of the source and drain voltages to account for the charge-sharing effect. The charge-sharing modeling approach predicts the classical threshold-voltage roll-off for short channels, as well as a decreased substrate effect. However, the so-called reverse short-channel effect (RSCE) introduces a threshold voltage roll-up for shorter channels. This effect can be noted since submicron technologies are used, and can become even more important in certain deep submicron technologies. Furthermore, a stronger dependence of the threshold voltage on drain-to-source voltage than predicted by the charge-sharing approach is sometimes required, which is introduced by a model for the drain-induced barrier lowering (DIBL). These different effects will mainly affect the pinch-off voltage, through the effective threshold voltage and effective substrate effect factor, which become a function of geometry and bias. Another effect noticeable on short-channel transistors is the weak inversion slope which can be strongly degraded for shortest channels, due to onset of punchthrough, requiring further flexibility in the model. The effect of the weak inversion slope degradation will be accounted for by changing the argument of the current interpolation function.

As in common practice, the offsets between drawn and effective device geometries are first introduced, using the two parameters D_L and D_W,

$$L_{eff} = L + D_L \quad W_{eff} = W + D_W \tag{24}$$

The channel length correction D_L will be treated as a constant for a given technology as usual in many other models. Some models consider also a bias-dependency of this parameter. However, considering both dependencies of series resistance and channel length with bias would lead to a non-solvable system of equations. Therefore, the bias dependency of series resistance, which will be retained, also accounts for eventual bias dependent effective channel length.

In the following, the models for the different effects are presented. Their fundamental form often corresponds to classical formulations also found in the literature. A number of similarities can for instance be noted with the PCIM model [18]. Mathematical conditioning is required to ensure continuity and correct behavior of the equations in all regions of operation; while this is necessary also elsewhere, this is particularly the case for the formulation of the short-channel effects. Therefore, no exhaustive presentation is made here; rather the basic forms are shown. Yet another effect affecting mainly short-channel devices is substrate current, generated through impact ionization of high energy carriers. Substrate current is added to the drain current, causing a degradation of the output conductance. This effect is also included in the model but will not be described here. While these models are reasonably simple, their combination has shown good accuracy in most situations, and also allows reasonably simple parameter extraction.

3.1 Velocity saturation and channel length modulation (CLM)

An increase in the field along the channel due to increasing drain-to-source voltage causes the velocity of carriers to saturate, resulting in a considerable reduction of drain current of short-channel transistors. A suitable velocity-field relationship has to be included in the model. Furthermore, as the device operates in saturation, the point in the channel where velocity saturation occurs moves towards the source, resulting in a degraded output conductance, which is referred to as channel-length modulation. Velocity saturation is accounted for in the effective mobility term as

$$\mu_{eff}' = \frac{\mu_{eff}}{1 + \dfrac{V_{DS}'}{E_C \cdot L_{eff}}} \tag{25}$$

The auxiliary function V_{DS}' ensures adequate behavior among conduction and saturation regions,

$$V_{DS}' = \begin{cases} V_D - V_S & for:\ V_D - V_S << V_{DSS} \\ V_{DSS} & for:\ V_D - V_S >> V_{DSS} \end{cases} \tag{26}$$

where the drain-to-source saturation voltage is expressed in strong inversion as

$$V_{DSS} \cong E_C \cdot L_{eff} \cdot \left[\sqrt{1 + \frac{2 \cdot (V_P - V_S)}{E_C \cdot L_{eff}}} - 1 \right] \tag{27}$$

and where the longitudinal critical field E_C is a temperature dependent model parameter.

To account for channel length modulation, the effective channel length in the specific current (11) is replaced by an equivalent channel length L_{eff}',

$$L_{eff}' = L_{eff} \cdot \left(1 - \frac{\Delta L}{L_{eff}}\right) \tag{28}$$

The term ΔL has a logarithmic dependence on $V_D - V_S$, according to a simplified pseudo two-dimensional analysis [19]

$$\Delta L = \lambda \cdot L_C \cdot \ln(1 + \frac{V_{DS}''}{E_C \cdot L_C}) \tag{29}$$

where λ is a model parameter. The characteristic length L_C is defined as a function of the oxide capacitance and junction depth X_J [19]

$$L_C = \sqrt{\frac{X_J \cdot \varepsilon_{si}}{C_{ox}'}} \tag{30}$$

The auxiliary function V_{DS}'' ensures that channel length modulation only occurs in saturation,

$$V_{DS}'' = \begin{cases} 0 & for: \ V_D - V_S << V_{DSS} \\ V_D - V_S - V_{DSS} & for: \ V_D - V_S >> V_{DSS} \end{cases} \tag{31}$$

Further details of the exact model formulation can be found in [10].

3.2 Charge-sharing and reverse short-channel effect (RSCE)

Normally, the threshold voltage decreases monotonically with decreasing the channel length [8], which is modeled using the conventional charge-sharing approach. However, in present submicron CMOS technologies, the threshold voltage initially increases when decreasing the channel length, reaches a maximum value, and then rolls off when the usual charge-sharing effect becomes dominant. The RSCE effect is often observed with n-channel devices, but may not be present in p-channel devices.

The charge-sharing principle [20] is based on geometrical considerations, for short-channel devices where the source and drain depletion regions overlap with the channel region. A reduction in the effective depletion charge controlled by the gate is predicted, depending on the extensions of the two depletion regions controlled by the source-bulk and drain-bulk junctions. The depletion depths $W_{S(D)}$ around the junctions depend on $V_{S(D)}$ as

$$W_{S(D)} = \zeta \cdot \sqrt{\psi_0 + V_{S(D)}} \quad with: \zeta = \sqrt{\frac{2 \cdot \varepsilon_{si}}{q \cdot N_{sub}}} \tag{32}$$

In narrow channel devices, the depletion region beneath the channel inversion region is not strictly confined laterally but fringes out. Therefore, an increased gate voltage is required to control the total depletion charge, resulting in an increased threshold voltage and substrate effect.

The charge-sharing effect for both short and narrow channels can be accounted for by introducing an effective substrate effect factor, which is formulated as follows [8][1][10]:

$$\gamma_{eff} = \gamma - g_L \cdot \left(\sqrt{\psi_0 + V_D} + \sqrt{\psi_0 + V_S}\right) + g_W \cdot \sqrt{\psi_0 + V_P} \tag{33}$$

where

$$g_L = \frac{\eta_L \cdot \varepsilon_{ox}}{C_{ox}'} \cdot L_{eff}^{-lex} \quad g_W = \frac{3 \cdot \eta_W \cdot \varepsilon_{ox}}{C_{ox}'} \cdot W_{eff}^{-wex} \tag{34}$$

The above relationship shows that the substrate effect factor is reduced for shorter channel length, while it is increased for narrow channel widths. Two empirical parameters η_L and η_W are used, related to the extension of the respective depletion regions. The exponent of the effective channel length usually takes a value of $lex = 1$, and similarly $wex = 1$ for effective channel width; the latter are handled as empirical model parameters. The charge-sharing effect is in general only weakly temperature dependent. γ_{eff} is then used in the equation of the pinch-off voltage (6) [1]:

$$V_{Peff} = V_G' - \psi_0 - \gamma_{eff} \cdot \left[\sqrt{V_G' + \left(\frac{\gamma_{eff}}{2}\right)^2} - \frac{\gamma_{eff}}{2}\right] \tag{35}$$

where $V_G' = V_G - V_{TO} + \psi_0 + \gamma \cdot \sqrt{\psi_0}$. The resulting effective slope factor is obtained as

$$n_{eff} = 1 + \frac{\gamma_{eff}}{2 \cdot \sqrt{\psi_0 + V_{P_{eff}}}} \tag{36}$$

Note that γ_{eff}, besides changing the substrate effect, also introduces a drop of the effective threshold voltage (which is referred to bulk) for short channels, while it is increased for narrow channels:

$$\Delta V_{T(CS)} \cong -\sqrt{\psi_0} \cdot \left[g_L \cdot \left(\sqrt{\psi_0 + V_D} + \sqrt{\psi_0 + V_S}\right) - g_W \cdot \sqrt{\psi_0}\right] \tag{37}$$

The reverse short-channel effect (RSCE) has been found to depend on the energy and dose of punchthrough implant and reoxidation time [21]. It is explained on the basis of non-uniform channel dopant distribution along the channel region due to transient enhanced diffusion of dopants. RSCE not only considerably affects device behavior, but also complicates MOST modeling due to its strong variability with channel length. Peaks of the threshold voltage of as much as $100mV$ or even more above the long-channel value V_{TO} can be observed. The location of the peak is variable as well, and it can even be situated at the shortest channel length, such that no threshold voltage roll-off is observed at all. Various models for the RSCE have been proposed, either as a correction of substrate doping [22], or as a correction on the threshold voltage [23]. They have the disadvantage to use CPU-expensive exponential terms, which also may introduce serious convergence problems during the related parameter extraction, due to the high sensitivity of exponential expressions.

A simple but reasonably accurate model of the change in threshold voltage, without use of exponentials, is derived from the original expression used by Arora [23]. It is given by [10][11]

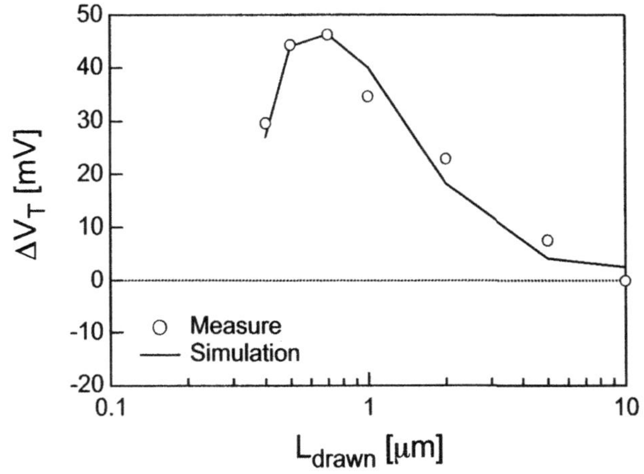

Figure 4. Variation of the threshold voltage due to RSCE as a function of the drawn channel length for n-channel devices of a $0.5\mu m$ technology.

$$\Delta V_{T(RSCE)} = \frac{2 \cdot Q_0}{C_{ox}'} \cdot \left[1 + \frac{1}{2} \cdot \left(\xi + \sqrt{\xi^2 + 4 \cdot \varepsilon_1^{\ 2}}\right)\right]^{-2} \tag{38}$$

where $\xi = C_1 \cdot (10 \cdot L_{eff} / L_K - 1)$, ε_1 and C_1 are two constants. (38) contains two parameters, related to the peak doping density at the source/drain ends, Q_0, and the characteristic length L_K over which the doping distribution is spread.

The threshold voltage variation due to RSCE given by (38) is combined with the threshold voltage reduction resulting from the charge-sharing model. The overall threshold voltage variation with respect to its long-channel value V_{TO} is plotted in Fig. 4 as a function of the drawn gate length in the case of a $0.5\mu m$technology. Fig. 4 shows that the RSCE is well modeled without the need for exponential functions, making the extraction of the related parameters Q_0 and L_K more robust.

Note that narrow-channel effects also may show a more complicated behavior than introduced so far in the model. Generally, a strong dependency on the isolation scheme is observed [17]. In particular, inverse narrow-width effects (INWE) have been observed, resulting in an analogous but opposite behavior as with the RSCE: threshold voltage may decrease for narrower channel widths [24], before it starts to increase when the usual narrow-channel effects start to dominate. Since these effects however can even be considerably more complicated, they are not introduced here.

3.3 Drain induced barrier lowering (DIBL)

In short-channel devices, and in particular for deep submicron technologies in use at present, an additional effect can be noted to have an important impact on the device characteristics: drain induced barrier lowering (DIBL) is commonly observed as a decrease of the threshold voltage for an increased drain bias (see e.g. [17]). DIBL models commonly introduce a bias- and channel-length-dependent threshold voltage variation. Various dependencies on drain-to-source voltage and channel length have been proposed. Many models use a linear relationship between threshold voltage shift and $V_D - V_S$, which is adequate at large $V_D - V_S$. Theoretical analysis and measurements show that this behavior may change considerably at lower $V_D - V_S$, where a stronger than linear relationship is observed. However, experimental results suggest that the linear relationship in combination with the charge-sharing model provides reasonable accuracy [11]:

$$\Delta V_{T(DIBL)} = -\sigma \cdot \left(V_D - V_S\right) = -\frac{\sigma_0 \cdot \varepsilon_{si}}{C_{ox}'} \cdot L_{eff}^{-slex} \cdot \left(V_D - V_S\right) \tag{39}$$

where σ_0 is the DIBL related model parameter. The formulation used here is comparable to the one used in [18]. The exponent for effective length dependence usually ranges from 2 to 3. To simplify computation and parameter extraction, it is linked to the exponent of the charge-sharing model by fixing its value to $slex = lex + 1$. In several modeling approaches, the effect of the threshold voltage shift is maintained only in weak inversion, while it is eliminated in strong inversion. Experimental results however suggest that maintaining the shift even in strong inversion does not adversely affect the simulated device behavior. The charge-sharing already predicts a lowering of the threshold voltage in the same direction as DIBL. The global threshold voltage model is now a result of the combined effects of the charge-sharing, RSCE and DIBL effects [17] with geometry and bias:

$$V_{Teff} = V_{TO} + \Delta V_{T(CS)} + \Delta V_{T(RSCE)} + \Delta V_{T(DIBL)} \tag{40}$$

In practice, the related parameters are strongly process-dependent.

3.4 Weak inversion slope degradation

The pinch-off voltage concept allows to accurately predict the weak inversion slope for long-channel devices. However, as noted earlier, short-channel devices can be affected by a reduced weak inversion slope, indicating that punchthrough is starting to occur. The behavior with channel length can however be strongly process-dependent. If no threshold voltage adjust implant is used, the weak inversion slope can be gradually degraded with shorter channel length, while a threshold voltage adjust implant can even improve the weak inversion slope for shorter channels, before it starts degrading for very short channels [25]. The improvement in weak inversion slope is explained on the basis of charge-sharing between gate and drain. Note that the charge-sharing model introduced before actually predicts an improved weak inversion slope for shorter channels due to the use of the effective slope factor n_{eff} (36). The weak inversion slope degradation calls for additional flexibility in the model, which is introduced through a modification of the argument of the drain current interpolation function $F(v)$ in the following manner:

$$F_\eta(v) = \eta_v{}^2 \cdot F\left(\frac{v}{\eta_v}\right) \tag{41}$$

Note that this change has a similar effect as if the temperature were changed when calculating U_T used in the normalized voltage $v \sim 1/U_T$ and in the specific current $I_S \sim U_T{}^2$. The resulting current in weak inversion is

$$I_D \cong I_S \cdot \eta_v{}^2 \cdot \exp(\frac{V_{Peff} - V_S}{\eta_v \cdot U_T}) \cong I_S \cdot \exp(\frac{V_G - V_{T_{eff}} - n_{eff} \cdot V_S}{\eta_v \cdot n_{eff} \cdot U_T}) \tag{42}$$

The factor $\eta_v{}^2$ ensures that the asymptotic behavior remains unchanged in strong inversion, while its influence in weak inversion is negligible compared to the effect of η_v in the argument of the exponential. Also note the similarity of this formulation with the one used in [18]. Here however the slope factor correction η_v is itself linked to L_{eff} through the charge-sharing and DIBL models:

$$\eta_v = 1 + \nu_L \cdot f(\gamma_{eff}, \sigma) \tag{43}$$

where the parameter ν_L allows the weak inversion slope to be adapted; the function $f(\gamma_{eff}, \sigma)$ accounts for the change in the slope factor with respect to the long-channel value due to the charge-sharing and DIBL models. This function also guarantees that the normalized transconductance-to-current ratio, for modulation from either the gate or the source, cannot exceed the physical limit of 1. The effective slope factor for short channel devices is therefore equal to $\eta_v \cdot n_{eff}$, while the subthreshold swing S_G becomes

$$S_G \cong 2.3 \cdot \eta_v \cdot n_{eff} \cdot U_T \tag{44}$$

For long channels, η_v tends to 1, making the influence on weak inversion slope vanish.

3.4 Gate voltage dependent series resistance

Series resistance is a critical parameter for process engineering, device modeling and circuit design of present submicron CMOS technologies using most often lightly doped drain (LDD) structures. In this context, the series resistance cannot be considered as constant and independent of the gate voltage. The resistivity of the gate overlapped accumulated LDD region will be affected by the gate voltage. Typically, a decreasing series resistance with increasing gate voltage is observed. Not accounting for this effect may lead to additional errors in modeling and of course in simulation. The variation of the series resistance with the gate voltage may become less important for advanced technologies such as $0.15\mu m$and $0.1\mu m$, to nearly constant values when the effective gate bias $V_{Geff} = V_G - V_{Teff}$ is increased [26. Nevertheless, even small variations of the resistance with the effective gate bias have a non-negligible effect on the extraction of both channel length reduction and series resistance with V_{Geff} [26], when the popular so-called resistance based method is used.

Various models of the bias dependent series resistance have been proposed [26][27]. Unfortunately, they are rather complicated and use computationally expensive exponential functions. The simple model for the bias dependent series resistance proposed here avoids the use of exponentials without noticeable loss of accuracy [11]

$$R_{S,Deff} = \frac{R_l}{W_{eff}} \cdot \left[1 - \left(\frac{1}{2} \cdot r + \sqrt{r^2 + 4 \cdot \varepsilon_2^{\ 2}}\right)\right] \tag{45}$$

where $r = S_{VK} \cdot (V_{Geff} / V_k - 1)$ and ε_2 is a constant. The parameters in (45) are V_k, S_{VK} and R_l, the diffusion resistivity at $V_{Geff} = 0$. Equation (45) is plotted versus the effective gate voltage in Fig. 5 and compared to measured data for a $0.35\mu m$technology [22]. The approximation error is well below 1% for the whole range of V_{Geff}.

Including the effect of series resistance explicitly in the expression for drain current has several advantages: it allows the gate voltage dependence to be introduced, whereas if external resistances are used, such a dependence is difficult to establish; also, the number of iterations when solving the circuits is reduced. Parameter extraction itself is simplified as well since solving for the extra nodes is avoided. The following relations are used as discussed in [28] :

$$\frac{I_D}{I_{D0}} \cong \frac{g_{mg}}{g_{mg0}} \cong \frac{g_{ms}}{g_{ms0}} \cong \frac{g_{md}}{g_{md0}} \cong \frac{1}{1 + g_{ms0} \cdot R_{Seff} + g_{md0} \cdot R_{Deff}} \tag{46}$$

where the subscript 0 denotes current or conductances calculated without series resistance. This relation shows that the impact of series resistance is greatest in conduction, where g_{md} is highest. In saturation, the series resistor on the drain side will have a negligible effect.

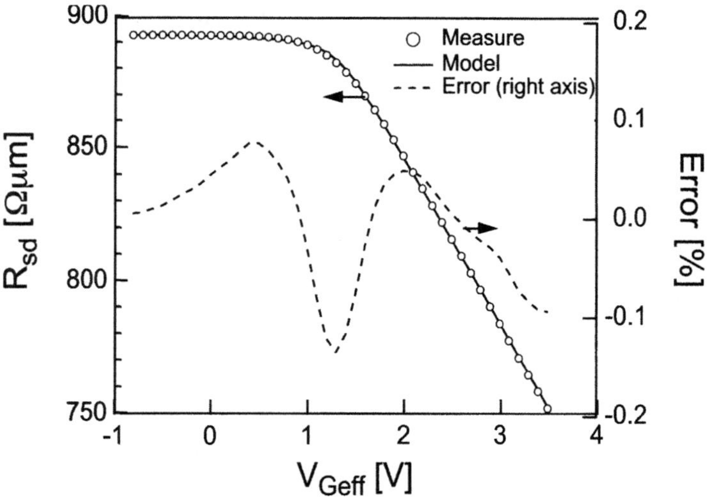

Figure 5. Gate bias dependence of the series resistance for an n-channel device of a $0.35\mu m$ CMOS technology [22].

This approach is more accurate than the usual scheme of accounting for series resistances which neglects the substrate bias conductance. Note however also that this approach, while it yields accurate results for DC, will have a negative impact on AC and transient analysis, especially for high frequency applications. The series resistance is therefore accounted for through constant external resistors independent of bias, and through an internally accounted bias dependent part using (46).

4. THE CHARGE AND THERMAL NOISE MODELS

4.1 Charges integration

The derivation of the quasi-static charge model is based on the fundamental relation existing between the inversion charge density Q_{inv} and the transconductance g_{ms} at point $x = 0$ in the channel where the channel voltage is equal to the source voltage V_S

$$-Q_{inv}'(V_S) = \frac{C_{ox}'}{\beta} \cdot g_{ms}(V_S) \tag{47}$$

Replacing g_{ms} by (15) leads to

$$-Q_{inv}'(x=0)=\frac{C_{ox}'}{\beta}\cdot\frac{I_F}{U_T}\cdot G(i_f)=2\cdot n\cdot U_T\cdot C_{ox}'\cdot i_f\cdot G(i_f) \tag{48}$$

The same relation exists at any point along the channel where the channel voltage is equal to

$$-Q_{inv}'(x)=\frac{C_{ox}'}{\beta}\cdot\frac{I_X}{U_T}\cdot G(i_x)=2\cdot n\cdot U_T\cdot C_{ox}'\cdot i_x\cdot G(i_x) \tag{49}$$

where $I_x = I_S \cdot i_x$ is the current variable, evaluated for $V_S = V_x$. Using the relationship $dx = -L/(i_f - i_r)\cdot di$ between the position and the normalized current [1], the total charge stored in the channel is then obtained by integrating (49) from the source, where i is equal to the forward normalized current i_f, to the drain where i is equal to the reverse normalized current i_r [1].

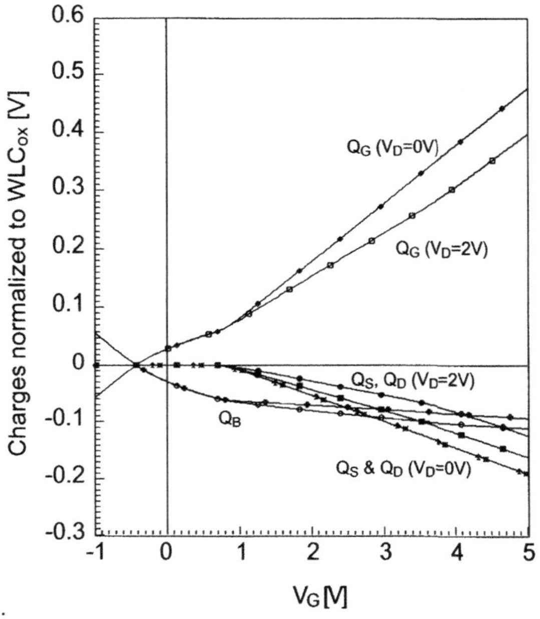

Figure 6. Node charges versus gate voltage for $V_S = 0V$ and for $V_D = 0V$ and $V_D = 2V$.

$$Q_{inv} \equiv W \cdot \int_0^L Q_{inv}'(x) \cdot dx = -\frac{W \cdot L}{i_f - i_r} \int_{i_f}^{i_r} Q_{inv}'(i) \cdot di \tag{50}$$

$$= -\frac{2 \cdot n \cdot U_T \cdot C_{ox}}{i_f - i_r} \int_{i_f}^{i_r} i \cdot G(i) \cdot di = -\frac{2 \cdot n \cdot U_T \cdot C_{ox}}{i_f - i_r} \int_{i_f}^{i_r} \left(\sqrt{\frac{1}{4} + i} - \frac{1}{2} \right) \cdot di$$

$$= -n \cdot U_T \cdot C_{ox} \cdot \left(\frac{4}{3} \cdot \frac{\chi_f^2 + \chi_f \cdot \chi_r + \chi_r^2}{\chi_f + \chi_r} \right)$$

where $C_{ox} = W \cdot L \cdot C_{ox}'$ and $\chi_{f(r)} = \sqrt{1/4 + i_{f(r)}}$. The charge associated to the drain node is computed using a linear charge partitioning while performing the integration:

$$Q_D \equiv W \cdot \int_0^L \frac{x}{L} \cdot Q_{inv}'(x) \cdot dx = -\frac{2 \cdot n \cdot U_T \cdot C_{ox}}{(i_f - i_r)^2} \int_{i_f}^{i_r} (i_f - i) \cdot i \cdot G(i) \cdot di$$

$$-n \cdot U_T \cdot C_{ox} \cdot \left(\frac{4}{15} \cdot \frac{2 \cdot \chi_f^3 + 4 \cdot \chi_f^2 \cdot \chi_r + 6 \cdot \chi_f \cdot \chi_r^2 + 3 \cdot \chi_r^3}{(\chi_f + \chi_r)^2} - \frac{1}{2} \right) \tag{51}$$

The charge associated to the source terminal is calculated as $Q_S = Q_{inv} - Q_D$ and the gate charge is given by $Q_G = Q_{inv} + Q_B + Q_{ox}$, where Q_B is the depletion charge given by [1]

$$Q_B = -C_{ox} \cdot \gamma \cdot \sqrt{\psi_0 + V_P} - \left(1 - \frac{1}{n}\right) \cdot Q_{inv} \tag{52}$$

and Q_{ox} is an eventual fixed oxide charge.

All the charges are plotted versus the gate voltage for two different drain voltages ($V_D = 0V$ and $V_D = 2V$) in Fig.6, which clearly demonstrates the continuity with respect to V_G and the symmetry of Q_S and Q_D with respect to source and drain sides.

Note that this charges model and the related transcapacitances model described in the next section, which are both used in [10], have been independently reported in [7] where they are obtained in a very similar manner, using the pinch-off voltage and normalized current concepts.

4.2 Transcapacitances model

The above formulation of the charge model allows charge conservation in transient simulation. The simulation algorithms require the partial derivatives of the charges, or transcapacitances, with respect to the terminal voltages to be formulated:

$$C_{XY} = \pm\frac{\partial Q_X}{\partial V_Y} \quad for \quad x,y = G,S,D,B \tag{53}$$

where the + sign is used in case $X = Y$ and the - sign otherwise. The corresponding expressions are not shown here for brevity. Some of the intrinsic capacitances are compared in Fig. 7 with measurements from a $0.25\mu m$ technology. Entirely symmetrical transcapacitances with respect to source and drain voltages are obtained, since the node charges are formulated in symmetric terms of i_f and i_r, as illustrated for the case $V_D = V_S$ in Fig. 7. The agreement of the model with measurements is excellent.

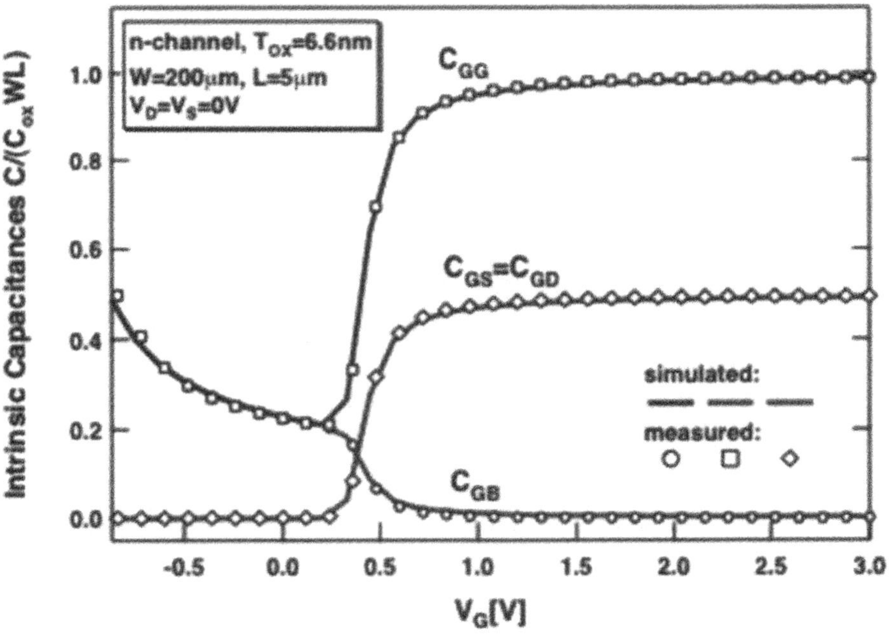

Figure 7. Measured and simulated transcapacitances versus gate voltage V_G, at $V_S = V_D = 0V$ from depletion to strong inversion, for a long n-channel device from a $0.25\mu m$ technology.

Note also that the pinch-off voltage V_P, used to calculate the basic variables used in the charges model, includes the short-channel effects used for the static model in the pinch-off voltage, namely charge-sharing, DIBL, and RSCE. Therefore, the most important short-channel effects are included in the dynamic charges/transcapacitances model. Additional short-channel effects, such as bias-dependent overlap capacitances, will be addressed elsewhere.

4.3 Noise model

The noise power spectral density can be expressed for each point of the channel as [1]

$$dS_{\Delta I_D} = 4 \cdot k \cdot T \cdot \frac{\mu(x)}{L^2} \cdot W \cdot Q_{inv}'(x) \cdot dx \tag{54}$$

The total power spectral density of the drain current fluctuations is then obtained through integration from source to drain, assuming constant mobility along the channel

$$S_{\Delta I_D} \cong 4 \cdot k \cdot T \cdot \frac{\mu_{eff}}{L_{eff}^2} \cdot W \cdot \int_0^L Q_{inv}'(x) \cdot dx$$

$$= 4 \cdot k \cdot T \cdot \frac{\mu_{eff}}{L_{eff}^2} \cdot \left| Q_{inv} \right| \tag{55}$$

however using the effective value of mobility resulting from (23) and effective channel length. Similarly as for the charges model, part of the short-channel effects are included through the pinch-off voltage. Since the expression (54) is general, the thermal noise model is valid from weak to strong inversion. Note that thermal noise does not vanish at $V_{DS} = 0V$, as is incorrectly predicted by the models used in early SPICE versions, where the thermal noise expression is proportional to g_{mg} which can be seen to be equal to zero in Table III at $V_{DS} = 0V$.

The thermal noise model is completed by a flicker or $1/f$ noise model, corresponding to expressions widely used with other MOST models:

$$S_{V_f} \sim 4 \cdot k \cdot T \cdot \frac{K_f}{W_{eff} \cdot L_{eff} \cdot f^{A_f}} \tag{56}$$

where the two parameters $A_f \cong 1$ and K_f are used. The parameter K_f can significantly vary with the processes; often, p-channel devices, in case they use buried channel, exhibit lower flicker noise than n-channel devices. Measurements show some bias dependency of the parameter K_f. For many design applications, this dependence can however be neglected.

5. MODEL APPLICATION AND EXPERIMENTAL RESULTS

5.1 The computer simulation model

The scalable model described previously has been implemented as a compact model in the circuit simulator ELDO. Coding of the entire set of model equations requires much more than just assembling the different pieces of the model. One of the important requirements for MOST models for circuit simulation is continuity of drain current and higher-order derivatives, among all operating regions, as well as outside of normal ranges of operation. Such conditions may occur while the simulator establishes operating points; therefore, mathematical conditioning is required that avoids functions to over- or underflow and ensures robustness. The drain current as well as the other model quantities are in principle formulated as single equations, conditioned such that they are valid in all regions of operation. Convenient smoothing functions are intensely used (see e.g. the set documented in [29]). However, much care is needed in choosing them so that fundamental physical behavior is not adversely affected.

The complete intrinsic model including series resistance requires a total of 23 process and DC parameters, which are described in Table IV. Two more parameters are required if vertical non-uniform doping is accounted for; three parameters are required to account for substrate current, for a total of 28 parameters. This number compares favorably with the more than 65 process and DC parameters required by the BSIM3v3 model [27][30] without counting its effective length and width sensitivity parameters. Temperature effects are formulated to affect the parameters V_{TO}, ψ_0, $K_p (= \mu_0 \cdot C_{ox})$, E_C, series resistance and substrate current, allowing good modeling of temperature behavior over the usual temperature range from 0-100°C. A non-quasistatic small-signal model is also provided, which uses first-order transadmittances [1], requiring no additional parameters.

The public-domain MOST model called 'EPFL-EKV v2.6' [10] (see also http://legwww.epfl.ch/ekv/), is available in most circuit simulators, among which ADS, ANTRIM-AMS, APLAC, ELDO, HSPICE, PSPICE, SABER, SMARTSPICE, SMASH, SPECTRE. This model is built using the same fundamental approach for the transconductance-to-current ratio as described here. The mobility model and short-channel effects are slightly simpler, using a reduced parameter set. Good adaptation for CMOS technologies in the submicron to quarter-micron range has been found and numerous analog ICs have been successfully designed.

This model also includes all the additional features as described in this section, making it suitable in particular for low-voltage, low-current applications in analog and mixed analog-digital circuit simulation.

5.2 Hierarchical model structure

The EKV model is hierarchically structured so that a designer can arbitrarily include or exclude particular physical effects, helping him to gain insight into the underlying physics of the MOS transistor and their impact on the circuit.

The minimum parameter set required would typically be `COX`, `KP`, `VTO`, `GAMMA` and `PHI`, while other parameters are disabled. This results in the fundamental "ideal" long-channel behavior, corresponding to the simple expressions used for hand-calculation in the asymptotic operating regions, but also including the transition regions of moderate inversion. Therefore a direct link between the hand calculation and the computer simulation models can be established, a unique feature of this MOST model. Other effects, as mobility degradation and short-channel effects, can then be gradually added and their impact on device characteristics and circuits studied. Such a procedure imposes a number of constraints on the model formulation, rarely met in other models: the different effects modeled should remain sufficiently independent of each other and the fundamental asymptotic behavior should remain correct.

5.3 Statistical circuit simulation including matching

Device mismatch, resulting from various random processes during device fabrication, can be observed between the parameters of equally designed devices [31]. In many analog or mixed analog-digital circuits, device mismatch is the main factor limiting their performance. The standard deviation of parameters for matched transistors is often observed to follow a dependence on geometry as [31]:

$$\sigma_P \approx \frac{A_P}{\sqrt{W_{eff} \cdot L_{eff}}} \tag{57}$$

Table 4: Intrinsic model parameters, symbols and units. Parameters for vertical non-uniform doping, substrate current, temperature effects and noise are not included. All parameters are available in the public domain version of the model, EPFL-EKV v2.6, except those marked (*).

Parameter	Symbol	Description	Units
COX	C_{ox}'	Gate oxide capacitance	F/m2
XJ	X_J	Source & Drain Junction depth	M
VTO	V_{TO}	Nominal threshold voltage	V
GAMMA	γ	Body effect factor	V1/2
PHI	ψ_0	Bulk Fermi potential (2x)	V
KP	K_P	Transconductance parameter	A/V2
E1 (E0)	E_1	1st order mobility reduction coefficient	1/V
E2 (*)	E_2	2nd order mobility reduction coefficient	1/V
UCRIT	E_C	Longitudinal critical field	V/m
LAMBDA	λ	Depletion length coefficient	-
DL	D_L	Channel length correction	M
LETA	η_L	Charge sharing coefficient	-
LEX (*)	lex	Length exponent for charge-sharing	-
SIGMA0 (*)	σ_0	DIBL coefficient	AsV/m2
NUL (*)	η_L	Short-channel slope factor parameter	AsV/m2
Q0	Q_0	RSCE peak charge density	AsV/m2
LK	L_K	RSCE characteristic length	M
DW	D_W	Channel width correction	M
WETA	η_W	Narrow width effect coefficient	-
WEX (*)	wex	Width exponent for charge-sharing	-
RL (*)	R_L	Diffusion resistivity	M
VK (*)	V_K	Characteristic gate voltage for resistivity	V
SVK (*)	S_{VK}	Resistivity coefficient	-

where A_P is the area proportionality constant for the parameter P. The statistical variation between matched devices will therefore be degraded as the area of the devices is decreased. Such dependencies can be observed over a considerable range of geometries, as long as identically laid out transistors are considered. The device mismatch is also degraded with increasing distance; in practice, this can often be neglected if distances between matched devices are kept minimal.

Statistical parameters, used in Monte Carlo simulation, need to be made available in the model equations to account for the geometry dependence of mismatch. With the exception of a few dedicated tools [32], MOST models in most circuit simulators almost never address this crucial need. Introducing statistical parameters at the model level allows a significant reduction of effort for the designer when optimizing analog layouts. Therefore, the *statistical parameters* A_{VTO}, A_{KP} and A_{γ} [31] are introduced, affecting the model parameters V_{TO}, K_P and γ in the following way [10] :

$$V_{TOa} = V_{TO} + A_{VTO} / \sqrt{W_{eff} \cdot L_{eff}} \tag{58}$$

$$K_{Pa} = K_P \cdot (1 + A_{KP} / \sqrt{W_{eff} \cdot L_{eff}}) \tag{59}$$

$$\gamma_a = \gamma + A_{\gamma} / \sqrt{W_{eff} \cdot L_{eff}} \tag{60}$$

For each matched MOST in the circuit, an individual parameter set V_{TOa}, K_{Pa} and γ_a is generated according to the standard deviation of the parameter and depending on the device geometry. The matching parameters are sometimes supplied by foundries, since the knowledge of these parameters is a key figure of merit of a technology. An area efficient method to obtain such parameters from high density integrated matrices of MOS transistors has recently been presented [33].

5.4 The pinch-off voltage measurement and parameter extraction method

The parameter extraction method proposed here follows the principles outlined in [34]. Of particular importance is parameter extraction from the pinch-off versus gate voltage characteristic measured at a constant current, from which the threshold voltage and doping related parameters are obtained [13][34], as well as charge-sharing and RSCE parameters. This method allows direct extraction of some of the parameters involved, making the extraction sequence considerably simpler, while other parameters are obtained by local optimization.

The pinch-off voltage versus gate voltage characteristic can be directly measured at the source of the transistor, by biasing it at a constant current [34]. The bias current is chosen such that the transistor is operating close to the middle of moderate inversion, equal to approximately half the specific current I_S [34]. This level of current is determined as follows: imposing the normalized voltage $v = 0$ in (17) , such that $V_P \cong V_S$, and solving for the normalized current, yields $i_f \cong I_D / I_S \cong 0.618$. Using the circuit in Fig. 8 a) , the pinch-off voltage is then measured at the source while sweeping the gate voltage over the bias range of interest. The characteristics depend on device sizes as seen in the measurements from a $0.7\mu m$ technology in Fig. 9 , showing typical behavior. For short-channel, a decreased substrate effect (increased slope of the V_P vs. V_G characteristic) is commonly observed compared to long-channel, while it is increased for narrow-channel. Similarly, the threshold voltage decreases for short channel and increases for narrow channel, unless RSCE or INWE are present.

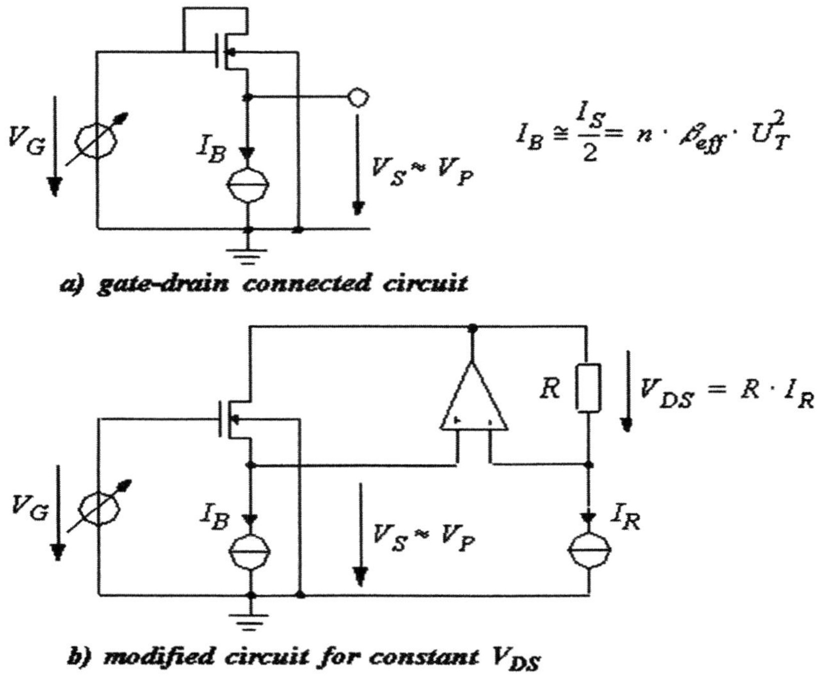

Figure 8. Circuit for pinch-off voltage V_P vs. V_G measurement with constant current bias in moderate inversion; a) simple measurement setup using gate-drain connected device, b) enhanced measurement setup using an OPAMP for measurement at constant V_{DS} .

In the simple measurement scheme of Fig. 8 a) using the gate-drain connection, the drain-to-source voltage varies only slowly, since the source voltage increases when sweeping the gate voltage. While this method has the advantage of simplicity, the measurement scheme of Fig. 8 b) can also be used, where an operational amplifier allows to control V_{DS} such that it remains strictly constant during the measurement. Instead, if available, an automatic feedback unit (AFU) can be used in instruments such as the HP4142 DC parameter analyzer, so that no dedicated measurement circuitry is required to perform such a measurement. A small drain-to-source voltage of $V_{DS} \cong 5 \cdot U_T$ is sufficient to ensure that the transistor is saturated, so that the measured source voltage $V_S \cong V_P$ can be interpreted as the pinch-off voltage.

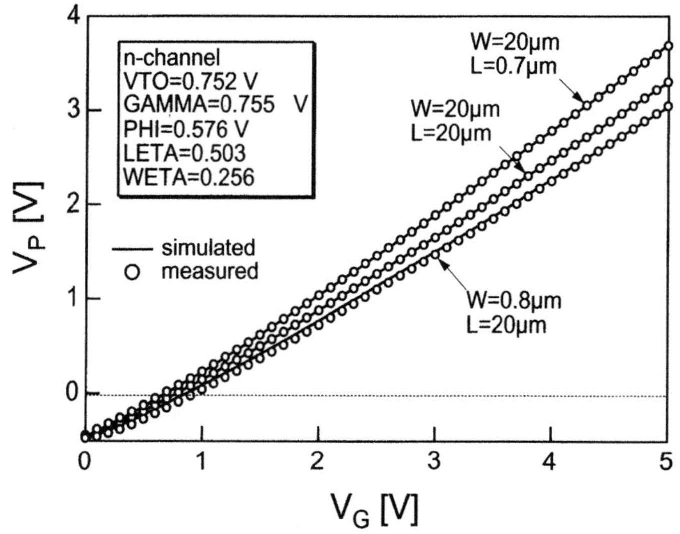

Figure 9. Measurement and parameter extraction from the V_P vs. V_G characteristics for long, short and narrow n-channel devices of a $0.7\mu m$ CMOS technology.

Long-channel parameters are determined from the measurement performed on a long and large transistor. The threshold voltage VTO is determined as the particular value of V_G corresponding to the $V_P \cong V_S = 0V$ crossing point [34]. GAMMA and PHI are extracted by fitting the analytical expression (6) of the long-channel pinch-off voltage to the measured characteristic. For the case where vertical non-uniform doping is considered, the related parameters can also be determined from the same measured characteristic [13].

The pinch-off voltage measured for transistors with varying channel length can then be used to extract the charge-sharing and RSCE related model parameters. Note that the current bias needs to be adapted for each geometry, so that the same inversion level is maintained for all devices. Similarly as for long-channel, the analytical expression of the pinch-off voltage V_{Peff}, including now the short-channel effects, is adapted to the measured characteristics. The charge-sharing related parameters `LETA` and `LEX` are determined first, by adjusting the *slope* of the V_P vs. V_G characteristics for various channel lengths. Typically three or more different channel lengths are used. The effective threshold voltage for each device geometry corresponds to the intersection points V_{Peff}; plotting $\Delta V_T = V_{Teff} - V_{TO}$ vs. L_{eff} results in the RSCE characteristic shown in Fig. 4 . The charge-sharing parameters already imply a roll-off of the threshold voltage. The parameters `Q0` and `LK` can be determined by matching the analytical expression V_{Teff} to the measured effective threshold voltage. The DIBL related parameter `SIGMA0` can be obtained if the pinch-off voltage measurement is performed at various V_{DS}, using if necessary the modified measurement circuit. A simple transformation of the pinch-off voltage characteristic $V_{Teff} = V_G - V_P$ allows to obtain the V_{TB} vs. V_P characteristic in a very simple manner [13]. An increased sensitivity is obtained when using the transform $V_{TB}(V_P)$ instead of $V_P(V_G)$ when extracting the related parameters.

Note that this method of characterization of the substrate effect, using a constant current bias in moderate inversion to measure the V_P vs. V_G characteristic, is not restricted to be used only with the EKV MOST model, but can also be applied with other MOST models. An important feature of this technique is that it allows to obtain threshold voltage and substrate effect from a single measured characteristic, as opposed to methods using extrapolated threshold voltage, which needs to be repeated for each back-bias. Another advantage is that this measurement technique is quite insensitive to variation of vertical field mobility and series resistance due to the low level of current that is applied.

An estimate of the specific current needs to be found for each transistor size, since the specific current is geometry- (and bias-) dependent. A reasonably simple scheme is the following: for a given gate voltage, i.e. a fixed pinch-off voltage, I_S can be determined, as indicated in Fig. 10, from the strong inversion slope of the $\sqrt{I_D}$ vs. V_S characteristic in saturation, which is an almost linear function except for the influence of series resistance which varies with current level. To minimize influence of mobility reduction due to vertical field and series resistance for short-channel transistors, V_G and V_D are chosen as low as possible, while maintaining the strong inversion and saturation condition. At any rate, a weak sensitivity of the measured pinch-off voltage characteristic and the extracted parameters with respect to the bias current chosen is observed [34].

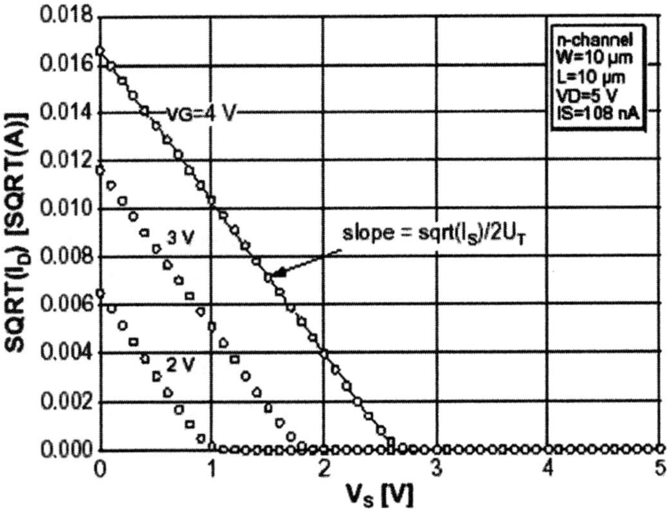

Figure 10. Measured $\sqrt{I_D}$ vs. V_S characteristic at various gate voltages V_G used to determine the specific current I_S.

5.5 Parameter extraction sequence

All process related model parameters can be initialized with known values from process technology, among which the gate oxide capacitance COX and the junction depth XJ. For increased precision, the gate oxide capacitance can be obtained from AC measurements. This is of importance when thin gate oxides are used, in particular for oxide thicknesses well below $t_{ox} \cong 10nm$. A DC parameter extraction sequence is proposed here which includes the above described method of the pinch-off voltage measurement and parameter extraction principle. As commonly required, a set of transistors with geometries covering the range of interest of W and L is used.

The first step includes extraction of geometrical offset (the parameter DL) and series resistance parameters, using e.g. the method in [26]. Long-channel threshold voltage and substrate effect related parameters are then obtained from the pinch-off voltage measurement method. From the same long-channel device, the mobility parameters KP, E1 and E2 can then be obtained from I_D vs. V_G at low V_{DS} using local optimization. Long-channel parameters are now fixed.

In the following, short-channel parameters are obtained. Threshold voltage and charge-sharing related parameters are obtained from the pinch-off voltage measured for each device, where a best fit over geometry and bias is achieved through the adjustment of the parameters `LETA`, `LEX`, `Q0`, `LK` and `SIGMA0`. The latter can also be obtained from I_D vs. V_G in weak inversion at different source biases V_S measured at low and high V_{DS} respectively. The parameter for weak inversion slope degradation `NU` is best obtained from the transconductance-to-current ratio g_{mg} / I_D in weak inversion. Short-channel parameters for velocity saturation (`UCRIT`) and channel length modulation (`LAMBDA`) can now be obtained from the output characteristics I_D and g_{md} vs. V_D in strong inversion at different gate voltages V_G. Substrate current parameters are best obtained from measured I_B vs. V_G characteristics at high V_D. In general, substrate current is only important for n-channel devices, for which g_{md} is degraded at high V_D.

Various methods exist to obtain the channel width correction DW, similarly to the channel length correction DL. It can also be simply obtained by adapting the I_D vs. V_G characteristics in strong inversion. The narrow-channel effect related parameters (WETA, WEX) are then obtained in a similar manner as short-channel parameters, using the pinch-off voltage method. Generally, reasonable results are obtained also for devices having both short and narrow channels, even though no particular modeling of the joined short- and narrow-channel effects has been introduced in the present model formulation.

Clearly, the pinch-off voltage measurement/extraction method considerably simplifies the parameter extraction. If this measurement capability is not at hand, it can be replaced by conventional techniques using

I-V curve fitting more extensively, at the cost of reduced efficiency and accuracy.

The above sequence can be refined and adapted to different technologies and/or particular operating regions if needed. It is sometimes necessary to reextract a given parameter if its value depends on another parameter to be subsequently extracted. The usage of mixed direct extraction and local optimization has shown to give a reasonable compromise between accuracy of model fit and efficiency. An automated procedure is in use when large amounts of data need to be gathered for statistical circuit simulation. The extraction method as described above has also shown to give good results with many different technologies, including deep submicron technologies.

5.6 Experimental results

In this subsection, a model validation, based on the simplified formulation of the public-domain model EPFL-EKV version 2.6 [10], is made. The scaling behavior for threshold voltage with channel length is illustrated for a $0.5\mu m$ technology in Fig. 4, where the simulated equivalent threshold voltage is shown to match the experimental data well. A small error of $6mV$ maximum for transistors with drawn channel lengths ranging from $0.4\mu m$ to $10\mu m$ is obtained [11].

For the same technology, the measured and simulated I_D vs. V_G characteristics in saturation and for different V_S are compared for three different channel lengths in Fig. 11 to Fig. 13 , for a long ($W = L = 10\mu m$), intermediate ($W = 10\mu m, L = 1\mu m$) and short channel device ($W = 10\mu m, L = 0.5\mu m$) respectively. The model fits well the experimental data from weak to strong inversion for all different channel lengths. Note also that the level of current corresponding to the specific current, approximately in the middle of moderate inversion is indicated for each geometry in the same figures.

Fig. 14 to Fig. 16 present the output characteristics I_D and g_{md} vs. V_D for different V_G for the same devices as above. The output conductance g_{md} both in conduction and saturation is well fitted for all geometries taking into account the small number of parameters that are used. These figures demonstrate the continuity of the model among all operating regimes and the good scaling behavior, since a single set of parameters has been used for all geometries.

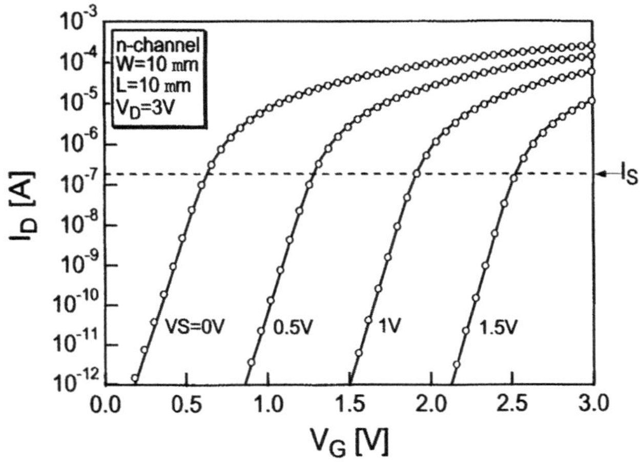

Figure 11. Transfer characteristics I_D vs. V_G of a $10\mu m/10\mu m$ n-channel device for different V_S ($0.5\mu m$ technology; symbols: measurement; lines: model). The level of the specific current I_S is indicated.

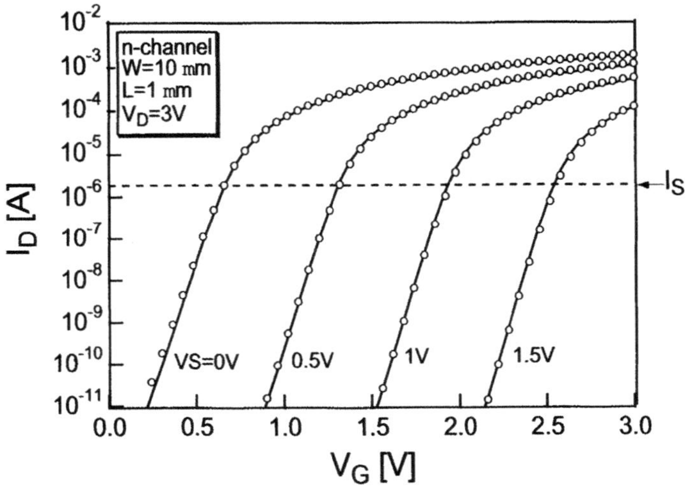

Figure 12. Transfer characteristics I_D vs. V_G of a $10\mu m/1\mu m$ n-channel device for different V_S ($0.5\mu m$ technology; symbols: measurement; lines: model). The level of the specific current I_S is indicated.

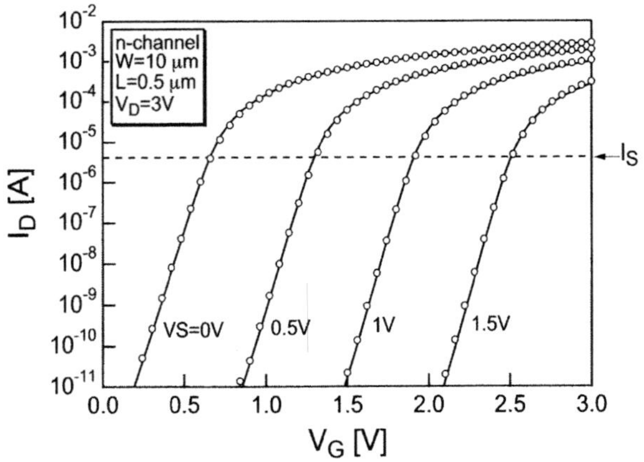

Figure 13. Transfer characteristics I_D vs. V_G of a $10\mu m/0.5\mu m$ n-channel device for different V_S ($0.5\mu m$ technology; symbols: measurement; lines: model). The level of the specific current I_S is indicated.

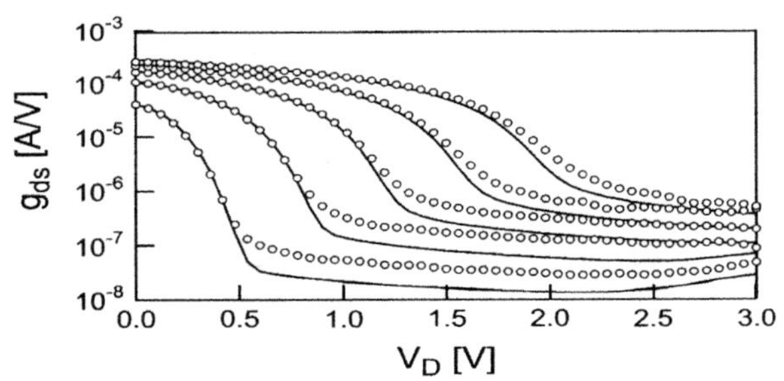

Figure 14. Output characteristics I_D and g_{md} vs. V_D at different V_G of a $10\mu m/10\mu m$ n-channel device ($0.5\mu m$ technology; symbols: measurement; lines: model).

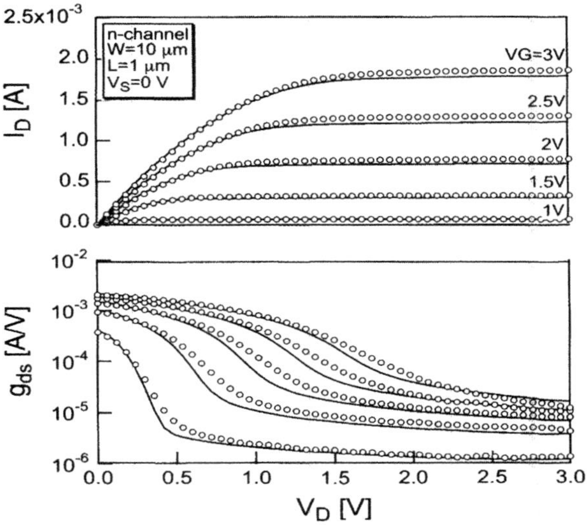

Figure 15. Output characteristics I_D and g_{md} vs. V_D at different V_G of a $10\mu m/1\mu m$ n-channel device ($0.5\mu m$ technology; symbols: measurement; lines: model).

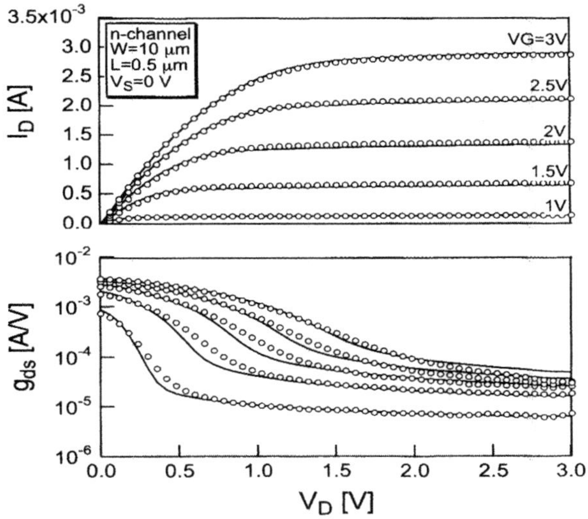

Figure 16. Output characteristics I_D and g_{md} vs. V_D at different V_G of a $10\mu m/0.5\mu m$ n-channel device ($0.5\mu m$ technology; symbols: measurement; lines: model).

6. CONCLUSIONS

An analytical compact charge-sheet model of the MOS transistor based on a physical description of the normalized transconductance-to-current ratio has been presented. The model uses the framework of the 'EKV' model approach, within which the pinch-off voltage, slope factor and normalized current are the principal model variables. It has been shown that the normalized transconductance-to-current ratio characteristic is independent of the technological parameters and therefore constitutes an ideal physical basis for the development of a complete compact circuit simulation model. The long-channel static model, the dynamic charge model as well as the thermal noise model, are all obtained by integration of the function describing the transconductance-to-current ratio. Therefore the present modeling approach is highly consistent and unified for all its aspects. The continuity of the ideal model equations and their n^{th}-order derivatives are therefore guaranteed. Hand-calculation expressions are developed, which are intensely used in analog circuit design.

The model for computer simulation includes all major physical effects present in submicron CMOS technology. Universal mobility dependence on vertical effective field is accounted for as well as effects of vertical non-uniform doping. Short-channel effects are formulated to complement the long-channel model: velocity saturation, channel-length modulation, source and drain charge-sharing, reverse short-channel effect, drain induced barrier lowering, weak inversion slope degradation and bias-dependent series resistance, making the model applicable to deep submicron CMOS technologies. A set of only 28 intrinsic model DC parameters is used, including vertical non-uniform doping, substrate current and bias dependent series resistance, comparing favorably to other models using typically more than 65 parameters. Thanks to the comparatively simple formulation, the model is also efficient and robust. To allow evaluation of geometry dependent device mismatch on analog circuits, statistical parameters are introduced for the main intrinsic model parameters for use in Monte-Carlo simulations.

The quality of circuit simulation results not only depends on the simulation model being used, but also critically depends on the quality of extracted device parameters. A parameter extraction method specially adapted to the model's structure and formulation, is therefore introduced. It is based on coupled direct extraction and local optimization techniques, allowing the sequence to be automated for the gathering of large amounts of data. In particular, an original measurement method, using constant current bias in moderate inversion, is used to determine threshold voltage and substrate effect related model parameters. Experimental results have been provided using measurements from CMOS technologies in the range of $0.7\mu m$ to $0.25\mu m$. The model's scalability is demonstrated for a $0.5\mu m$ standard CMOS technology, using a single parameter set.

The model presented in this chapter, in the version available in public domain, called EPFL-EKV v2.6, is used in a variety of contexts including deep submicron analog design. The basic model formulation is obtained from clear physical concepts and the model is well adapted to a large range of MOS technologies. Due to the model's strong link to circuit design practice, it also facilitates the portability and reuse of analog and mixed analog-digital ICs.

7. ACKNOWLEDGMENTS

It is a pleasure to the authors to acknowledge Wladek Grabinski for experimental support and Gerson A. S. Machado for numerous suggestions regarding the manuscript.

8. REFERENCES

[1] C. C. Enz, F. Krummenacher, E. A. Vittoz, "An Analytical MOS Transistor Model Valid in All Regions of Operation and Dedicated to Low-Voltage and Low-Current Applications," special issue of the Analog Integrated Circuits and Signal Processing Journal on Low-Voltage and Low-Power Design, vol. 8, pp. 83-114, July 1995.

[2] D. Foty, "MOSFET Modeling with SPICE, Principles and Practice", Prentice-Hall, 1997.

[3] C. C. Enz, E. A. Vittoz, "CMOS Low-Power Analog Circuit Design", Tutorial of ISCAS'96, Atlanta, USA, 1996.

[4] G. A. S. Machado, C. C. Enz, M. Bucher, "Estimating Key Parameters in the EKV MOST Model for Analogue Design and Simulation", Proc. IEEE Int. Symp. Circuits Syst., pp. 1588-1591, Seattle, Washington 1995.

[5] H. Oguey, S. Cserveny, "Modèle du transistor MOS valable dans un grand domaine de courants," Bull. SEV/VSE, Feb. 1982.

[6] G. Galup-Montoro, M. Schneider, S. Acosta, R. Pinto, *Proc. Brazilian Microel. Conf. SBMICRO'96*, 1996.

[7] A. I. A. Cunha, O. C Gouveia-Filho, M. C. Schneider, C. Galup-Montoro, "A Current Based Model for the MOS Transistor", Proc. IEEE Int. Symp. Circuits Syst., Vol. 3, pp. 1608-1611, June 1997.

[8] Y. P. Tsividis, "Operation and Modelling of the MOS Transistor", McGraw-Hill, 1987.

[9] C. A. Mead, "Scaling of MOS Technology to Submicrometer Feature Sizes", Analog Integrated Circuits and Signal Processing Journal , Vol 6, pp. 9-25, 1994.

[10] M. Bucher, C. Lallement, C. Enz, F. Théodoloz, F. Krummenacher, "The EPFL-EKV MOSFET Model Equations for Simulation, Model Version 2.6", Technical Report, Electronics Laboratories, Swiss Federal Institute of Technology (EPFL), Lausanne, Switzerland, June, 1997.

[11] M. Bucher, C. Lallement, C. Enz, F. Théodoloz, F. Krummenacher, "Scalable GM/I Based MOSFET Model", Proc. Int. Semicond. Device Research Symp., pp. 615-618, Charlottesville, VA, December 10-13, 1997 .

[12] C. C. Enz, "MOS Transistor Modeling Dedicated to Low-Current and Low-Voltage Analog Circuit Design and Simulation," in Low-power HF Microelectronics: A Unified Approach, Ed. by G. Machado, IEE Book Publishing, Ch. 7, 1996, pp. 247-299, ISBN 0 85296 874 4.

[13] C. Lallement, M. Bucher, C. C. Enz, "Modelling and Characterization of the Non-Uniform Substrate Doping," Solid State Electron., Vol. 41, No. 12, pp. 1857-1861, 1997.

[14] T. Hori, "Gate Dielectrics and MOS ULSIs, Principles, Technologies, and Applications", Springer Series in Electronics and Photonics 34, Springer, 1997.

[15] S. C. Sun, J. D. Plummer, "Electron mobility in inversion and accumulation layers on thermally oxidized silicon surface", Trans. IEEE Electron Dev., ED-27, pp. 1497-1508, 1980.

[16] N. D. Arora, G. Sh. Gildenblat, "A semi-empirical model of the MOSFET Inversion Layer for Low-Temperature operation", Trans. IEEE Electron Dev., ED-34, pp. 89-93, 1987.

[17] N. Arora, "MOSFET Models for VLSI Circuit Simulation - Theory and Practice", Computational Microelectronics, Springer Verlag, Wien New York, 1993.

[18] N. D. Arora, R. Rios, C.-L. Huang, K. Raol, "PCIM: A Physically Based Continuous Short-Channel IGFET Model for Circuit Simulation", Trans. IEEE Electron Dev., ED-41, No. 6, pp. 988-997, 1994.

[19] F. M. Klaassen, R. D. A. Velghe, Proc. ESSDERC 89, pp. 418-422, Springer, 1989.

[20] L. D. Yau, "A Simple Theory to predict the Threshold Voltage of Short-channel IGFETs", Solid State Electron., Vol. 17, pp. 1059-1063, 1974.

[21] M. Orlowski, C. Mazuré, F. Lau, "Submicron Short Channel Effects due to Gate Reoxidation Induced Lateral Interstitial Diffusion", IEEE IEDM Tech. Dig., pp. 632, 1987.

[22] H. Brut, "Contribution à la modélisation et à l'extraction des paramètres de tension de seuil, de résistance série et de réduction de longueur dans les transistors MOS submicroniques," Ph. D. thesis (in French), Institut National Polytechnique de Grenoble, December 1996.

[23] N. D. Arora, M. Sharma,"Modeling the Anomalous Threshold Voltage Behavior of Submicron MOSFETs", IEEE Electron Dev. Lett., EDL-13, pp. 92-94, 1992.

[24] L. A. Akers, M. Sugino, J. M. Ford, "Characterization of Inverse-Narrow-Width Effect", Trans. IEEE Electron Dev., ED-34, pp. 2476-2484, 1987.

[25] N. G. Tarr, D. J. Walkey, M. B. Rowlandson, S. B. Hewitt, T. W. MacElwee, "Short-Channel Effects on MOSFET Subthreshold Swing", Solid State Electron., Vol. 38, No. 3, pp. 697-701, 1995.

[26] H. Brut, A. Juge, G. Ghibaudo, "New Approach for the Extraction of Gate Voltage Dependent Series Resistance and Channel Length in CMOS Transistors", Proc. IEEE Int. Conf. on Microel. Test Structures, Monterey, USA, pp. 188-193, March 1997.

[27] BSIM3, Version 3.1 Manual, Department of Electrical Engineering and Computer Science, University of California, Berkeley, CA 94720, 1996.

[28] S. Cserveny, "Relationship between Measured and Intrinsic Transconductances of MOSFETs", Trans. IEEE Electron Dev., ED-37, no. 11, pp. 2413-2414, 1990.

[29] R. M. D. A. Velghe, D. B. M. Klaassen, F. M. Klaassen, "MOS Model 9", Unclassified Report NL-UR 003/94, Philips Electronics N.V. 1994.

[30] Y. Cheng, M.-C. Jeng, Z. Liu, J. Huang, M. Chan, K. Chen, P. K. Ko, C. Hu, "A Physical and Scalable I-V Model in BSIM3v3 for Analog/Digital Circuit Simulation", Trans. IEEE Electron Dev.,, Vol. 44, No. 2, pp. 277-287, February 1997.

[31] M. J. M. Pelgrom, A. C. J. Duinmaijer, A. P. G. Welbers, "Matching Properties of MOS Transistors", IEEE Journ. of Solid-State Circuits, Vol. SC-24, pp. 1433-1439, 1989.

[32] R. Weissenfels, J. Oehm, K. Schumacher, "YIELD: High Performance Analog Circuit Design with Regard to Statistical Aspects", Proc. 18th IEEE Europ. Solid-State Circ. Conf., pp.167-170, 1992.

[33] L. Portmann, C. Lallement, F. Krummenacher, "A High Density Integrated Test Matrix of MOS Transistors for Matching Study", Proc. IEEE Int. Conf. on Microel. Test Structures, Vol. 11, pp. 19-24, Kanazawa, Japan, March 23-26, 1998.

[34] M. Bucher, C. Lallement and C. C. Enz, "An Efficient Parameter Extraction Methodology for the EKV MOST Model," Proc. IEEE Int. Conf. on Microel. Test Structures, Vol. 9, pp. 145-150, March, 1996.

Chapter 4

Efficient Statistical Modeling for Circuit Simulation

C. C. McAndrew
Motorola, Inc., 2100 East Elliot Road, Tempe AZ, 85284 U.S.A. PH:(602)413-3982 FAX:(602)413-5343, mcandrew@sst.sps.mot.com

Key words: Statistical modeling, SPICE modeling, compact modeling, case files, skew files, backward propagation of variance, Monte Carlo circuit simulation.

Abstract: Statistical SPICE modeling is necessary for low risk IC design. Here existing approaches to statistical modeling are reviewed, and their limitations are discussed. A four level hierarchy of IC manufacturing variations is presented. Using physically based process and geometry level modeling, sensitivity analysis, and propagation of variance, it is shown how statistical models can be accurately and efficiently derived from the statistical distributions of key device electrical performances, as measured on manufacturing lines. The procedure runs in minutes of am engineering workstation, and guarantees accurate modeling of manufacturing variations.

1. INTRODUCTION

Modern business pressures mandate that integrated circuits (ICs) should be high yield, should work on the first pass through manufacturing, and should be as small as possible. This means that IC design methods are based on simulation, and that simulation must be based on models that accurately represent the electrical behavior of the components and parasitics that comprise the IC. In particular, IC designs should work within specifications over the statistical variations that are inherent in IC manufacturing technologies. Therefore, an important aspect of modeling for IC design is to provide statistical models for simulation.

Analog IC design is commonly done with a SPICE type simulator (Nagel, 1975). The models within such simulators are typically algebraic relations for branch currents $\boldsymbol{I}(\boldsymbol{V})$ and charges $\boldsymbol{Q}(\boldsymbol{V})$ as functions of branch voltages $\mathbf{V}$, and are termed compact models. Digital IC design is commonly based on synthesis, which uses timing models for digital cells. These timing models can be derived from SPICE simulations of the digital cells. Therefore here statistical modeling is addressed for SPICE compact models, for statistical simulation with a SPICE type simulator.

Several techniques for statistical modeling are used at present. These are reviewed and the drawbacks of each are detailed. In particular, it will be shown that existing techniques are inaccurate and/or inefficient.

To better conceptualize how statistical variations in IC manufacturing technologies are manifest, a four level hierarchy of manufacturing variations is used. This leads to modeling at the level of process and geometry parameters, rather than SPICE model parameters, as the best way to attack statistical modeling. Sensitivity analysis and propagation of variance, applied to process and geometry level models, are shown to lead to systems of equations that are easily solved to yield accurate statistical models.

The approach presented here is applicable to statistical modeling of any type of device. However, the examples presented will concentrate on MOSFET modeling. Applications for resistors and bipolar junction trans

istors (BJTs) will be touched upon, and more details of statistical BJT modeling are given elsewhere (McAndrew, 1997).

One fundamental principle of statistical modeling is that it must be based on sound statistical data. The best source of these data is process control (PC) measurements made on manufacturing lines, also called electrical test (E-test) data, class probe data, or final in-process test data. The statistical modeling approach presented here is based on PC data. Reliable PC data are not available until some time after a technology has been in manufacture, and the statistical spreads of PC data typically tighten over time as manufacturing control strategies improve. However, at any point in the life cycle of a manufacturing technology the statistics of the PC data can be defined by a combination of measurements, engineering experience, and expectations about equipment improvements, and this provides the necessary PC statistics to enable statistical SPICE models to be generated.

The term "modeling" is used in the context of SPICE simulation both to denote the development of equations that comprise a compact model, and to mean the extraction of the parameters of the compact model. The latter is often called "characterization". Here the term modeling will refer to both the formulation and characterization $I\ V(\ )\ Q\ V(\ )\ V$ procedures. The term "model file" denotes a SPICE .MODEL card.

It is assumed that statistical variations in IC manufacturing technologies can be characterized by normal (Gaussian) distributions. Also, it is assumed that these statistical variations are not large, so variations in measures of device electrical performance (i.e. PC data) can be modeled as linear functions of variations in the process parameters that are the root cause of manufacturing fluctuations. For measures of circuit performance this is not valid in general, as circuit behavior over the space of manufacturing variations can be highly nonlinear, even nonmonotonic (Ogrodzki, 1980). However the assumption of linearity is reasonable for PC data.

A result that will be used here is that if is normally distributed with mean and variance σ^2_1, i.e.

$$x_1 \sim N(\mu_1, \sigma_1^2), \tag{1}$$

and,

$$x_2 \sim N(\mu_2, \sigma_2^2), \tag{2}$$

then

$$y = a_1x_1 + a_2x_2 \sim N(a_1\mu_1 + a_2\mu_2, a_1^2\sigma_1^2 + a_2^2\sigma_2^2). \tag{3}$$

The mapping of the variances of x_1 and x_2 into the variance of is y termed the propagation of variance. For the multidimensional, correlated case, if

$$\boldsymbol{x} = \begin{bmatrix} x_1 & x_2 & \dots & x_n \end{bmatrix}^T \sim N(\mu, C), \tag{4}$$

where μ is the vector of the means of $\boldsymbol{x}$, $\boldsymbol{C}$ is the variance/covariance matrix of $\boldsymbol{x}$, and the superscript T means the vector transpose operation, then (Papoulis, 1991)

$$y = \boldsymbol{a}^T\boldsymbol{x} = \begin{bmatrix} a_1 & a_2 & \dots & a_n \end{bmatrix}\begin{bmatrix} x_1 & x_2 & \dots & x_n \end{bmatrix}^T \sim N(\boldsymbol{a}^T\mu, \boldsymbol{a}^T\boldsymbol{C}\boldsymbol{a}). \tag{5}$$

2. CLASSIFICATION OF STATISTICAL MODELS

There are several types of statistical models and statistical simulation techniques, and often the term "statistical modeling" is used without distinction as to which type is being addressed. For example, case (or skew) model files are used in an attempt to verify the manufacturability of an IC, and Monte Carlo type simulation techniques are used for more accurate yield analysis.

Here three different types of statistical models are explicitly identified and considered separately:

Distributional statistical models, characterized by means and variances and used for Monte Carlo type simulations;

Specific case statistical models that give extreme values for a specific measure of device performance; and

Generic case statistical models that give specified variations in key device electrical performances.

The case files commonly used for IC design are the last of these. However, generic case statistical models cannot guarantee accurate modeling of the variation of all measures of circuit performance for all circuits (topologies, biases, device sizing, etc.). For digital CMOS, generic case files are reasonable. This is because most digital circuits and cells are similar in terms of their topology (i.e. their "look and feel"), use minimum channel length devices, and have two important measures of circuit performance, speed and power, that are highly correlated. This means that generic case files that bracket the manufacturing distribution of the saturated drain current I_{dsat} (I_d and V_{sb} 0 = and $V_{ds} = V_{gs} = V_{DD}$ where V_{DD} is the supply voltage for the technology) for minimum channel length devices do a reasonable job of modeling the ranges of speed and power for most digital CMOS circuits. But for analog CMOS circuits, and nontypical digital CMOS circuits such as Schmitt triggers, the plethora of circuit topologies, measures of circuit performance, and device geometries and biases are such that generic case files that embody variations in I_{dsat} do not accurately model $\pm 3\sigma$ manufacturing variations in all measures of circuit performance for all circuits.

Techniques are presented here to accurately and efficiently generate each of the above types of statistical models.

3. HIERARCHY OF STATISTICAL VARIATIONS

Table 1 shows a four level hierarchy that is a useful conceptual tool for viewing statistical variations in IC manufacturing technologies. Modeling at the level of process inputs $\boldsymbol{i}$ is not feasible for circuit simulation. Although modeling at the level of SPICE model parameters $\boldsymbol{s}$ is the most common way to approach statistical modeling, there are many more SPICE parameters than process parameters $\boldsymbol{p}$, they can be highly correlated (especially for BJTs), and they can be poorly behaved statistically because of numerical "noise" in characterization. Because there are many fewer $\boldsymbol{p}$ than $\boldsymbol{s}$, and they are the physical cause of statistical fluctuations in manufacturing and are generally uncorrelated, it is apparent that statistical modeling for circuit simulation is best done at the level of process parameters. This is relatively straightforward for resistor and MOSFET models, which have a strong physical basis and are formulated in terms of process parameters, and Davis (1989) showed how to this for BJTs.

The device performances ***e*** will be used here as the basis for generating statistical models, and these are the PC data measured routinely on IC manufacturing lines. Rigorous mathematical criteria will be presented below to determine whether a particular set of ***e*** are sufficient for statistical characterization of a specific type of device. However, one general principle that should be followed is that the method of measuring ***e*** should be as direct as possible. The greater the amount and complexity

Table 1. Four level hierarchy for statistical variations

Level	Examples	Number	Comment
process inputs ***i***	implant doses and energies anneal ramps, times, temperatures oxidation ramps, times, temperatures	100's	Uncorrelated
Process parameters ***p***	R: ρ_S, Δ_{Lr}, Δ_{Wr} MOS: N_b, T_{ox}, V_{fb}, Δ_L, Δ_W BJT: ρ_{sbe}, N_{epi}, Δ_{Le}, J_{bei}, ρ_{sb}	10's	Nearly Uncorrelated
SPICE parameters ***s***	R: ρ_s, Δ_{Lr}, Δ_{Wr} MOS: Nb, Tox, Vth, ΔL, ΔW BJT: *IS, BF, CJE, VAF, RB, IKF*	100's to 1000's	Highly Correlated, poor statistically
Device/circuit performances ***e***	Device: R, V_{th0}, I_{dsat}, g_o, I_c, β Circuit: τ_p, P_{dis}, V_{ol}, ϕ_m, P_{1DB}	∞	

of the manipulations of raw measurements to arrive at some element e_i of ***e*** the greater is the chance of an error in calculationg e_i, either in the measurement procedure itself or in its duplication in simulation from models. In addition, some device performance measures derived from sophisticated extrapolation techniques can have significant differences between measured and modeled values, even when the model reasonably accurately represents normal device $\boldsymbol{I}$ ($\boldsymbol{V}$) behavior. The simpler and more direct the measurement techniques for ***e*** the more suitable and useful they are for the purpose of statistical modeling.

4. PROCESS AND GEOMETRY LEVEL MODELING

Statistical modeling at the level of process parameters implies that mappings must be defined between ***p*** and ***s***. Variations in device dimensions are key components of statistical variations in IC manufacturing technologies, therefore proper statistical modeling also must be based on models that depend on geometry $\boldsymbol{g}$. MOSFET models are typically based on process and geometry parameters. Resistor models also are typically based on the design length and width L_r and W_r, and the process parameters sheet resistance ρ_S and the differences between design and effective electrical length and width, Δ_{Lr} and Δ_{Wr}.

$$R = \rho_s \frac{L_r + \Delta_{Lr}}{W_r + \Delta_{Wr}} \tag{6}$$

(additional end resistances can also be included). However, BJT models are usually defined with parameters that are highly correlated, and that depend on both process and geometry parameters. Therefore the first step in statistical modeling for BJTs is to define the process and geometry mappings. This can be done using physical analysis, process and device simulation, and measurement (Davis, 1989).

The quantities such as Δ_L and Δ_W, that model the difference between the design and effective electrical dimensions of a device, are considered here as the process parameters that model the statistical variations in geometry.

5. EXISTING STATISTICAL MODELING APPROACHES

Several approaches have been used for statistical modeling. The most common of these are summarized and reviewed in this section.

5.1 SPICE model parameter perturbation

This is the easiest method of generating statistical models. A SPICE .MODEL card is extracted from data, and case files are generated by introducing perturbations in some of the model parameters.

This procedure is simple and fast. However, it has some severe drawbacks. MOSFET and resistor models are generally based on process parameters, and so perturbing these can give reasonable variations in modeled device electrical behavior. But there are correlations between SPICE model parameters, particularly for BJT models, and erroneous modeling of device behavior can result if these are not accounted for properly. In addition, this approach provides only case models, not distributional statistical models, and as discussed in section 5.3 cannot guarantee the accuracy of the modeling.

5.2 Extreme case data

Another approach to statistical modeling is to obtain extreme case data, either from split manufacturing lots or from physical simulations of manufacturing extremes (Prendergast, 1985), and then extract SPICE .MODEL cards from each set of data.

As with simple SPICE model parameter perturbations, this approach provides only case files and not distributional models suitable for Monte Carlo type statistical simulations. However, unlike direct SPICE model parameter perturbation this approach multiplies the characterization effort by the number of case files generated, which is inefficient. More important, split lots or physical simulation perturbations are based on estimates of variations in a few key process parameters, and are not guaranteed to accurately bracket actual manufacturing variations. The accuracy of split lot data is further confounded by the uncontrollable variations in manufacturing even with control settings set to expected manufacturing extremes. Techniques for selecting extreme case lots for statistical modeling have been reported (Chen, 1995), however these require significant manufacturing and measurement effort, are *ad hoc*, and do not guarantee accurate bracketing of all key device electrical parameter distributions.

5.3 Forward propagation of variance

Given a process and geometry level model, it is apparent that statistical modeling can be done by directly measuring the statistics of the process level model parameters (which include lateral geometry variations), and then basing statistical SPICE models on these variations. This is probably the most commonly used method to generate statistical SPICE models. It is well known for MOSFETs (Cox, 1985), and was pioneered by Davis (1989) for BJTs, by building process level models on top of the SPICE Gummel-Poon (SGP) model.

Unlike the approaches described in sections 5.1 and 5.2, this approach can provide both generic case and distributional statistical models. It can also be implemented efficiently, as the statistical information on $\boldsymbol{p}$ can be obtained provided it is measured as part of the PC data, although development and verification of the process parameter to SPICE model parameter mappings can require significant effort, especially for BJTs. Besides the inefficiency incurred in setting up the models, this approach has three other major drawbacks.

First, when case files are generated it is typically done by introducing specified variations, often $\pm 3\sigma$, in the process parameters. Not only does this generate excessively pessimistic case files (adding $\pm 3\sigma$ variations in each of

n independent variables results in a $\pm 3\sqrt{n}\sigma$ total variation), but it ignores the sensitivity of the device behavior to variations in the process parameters. For example, consider a MOS technology in which oxide thickness T_{ox} and effective channel length variation $\Delta_L = L_g - L_{eff}$ are the dominant sources of manufacturing fluctuations. For a circuit composed of long channel devices the $\pm 3\sigma$ variations in measures of circuit performance can be modeled by introducing a variation $\pm 3\sigma$ in T_{ox}. However, for a circuit composed of short channel devices both T_{ox} and Δ_L affect circuit performance, and therefore the points in process parameter space that give $\pm 3\sigma$ variations in T_{ox} circuit performance will have less than $\pm 3\sigma$ individual variations in T_{ox} and Δ_L . This is a generic problem with introducing fixed perturbations in process parameter space to generate case files. The solution to this problem is presented in section 8 below.

Second, it is not always easy to directly measure all process parameters. For example, epi doping and lateral PNP basewidth. Direct measurement can entail special structures and complex and time consuming data acquisition and manipulation, which are undesirable as this increases manufacturing cost because the test time to acquire the PC data increases. As noted in section 3, the PC tests should be as simple as possible, and should be made on standard devices and not on special test structures.

Third, and most important, the mappings both from the process parameters to the SPICE parameters, and from the SPICE parameters to the device and circuit performances, are approximations, and are different for different models. Therefore, directly feeding in perturbations or distributions in process parameters, which can be termed forward propagation of variance, fundamentally does not guarantee accurate modeling of device or circuit performance perturbations or distributions. Figure 1 shows this. In other words, because different models are different, feeding the same process parameter statistical distributions into different models will give different predicted distributions of device and circuit performances, which is clearly incorrect. In addition, there is often more than one method to determine a particular process parameter, MOSFET Δ_L being the most obvious example, and each method can give a different value. If these different values are directly fed into a model they will give different predictions of the distributions of the electrical behavior of the device, which is again clearly incorrect. The goal of statistical modeling is to accurately model the distributions of ***e***, not of ***p***, and forward propagation of variance does not guarantee this.

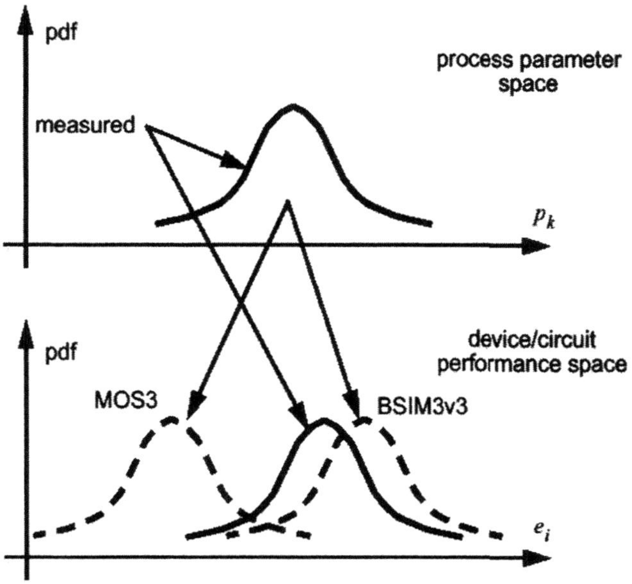

Figure 1. Errors involved in forward propagation of variance.

5.4 Numerical data fitting

Recently, standard numerical modeling techniques have been applied to the problem of statistical modeling, in the form of principle components analysis (PCA, Power 1994) and response surface modeling (RSM, Power 1997). SPICE model parameters and PC data are determined for die samples that represent the expected range of manufacturing variations, and then either PCA is applied to find a reduced dimension modeling space or numerical models between the PC parameters and the SPICE model parameters are derived from the data.

These numerical approaches can provide distributional statistical models, and can also provide case files, although as noted in section 5.3 if $\pm 3\sigma$ variations are made in all variables in the reduced modeling space then there is no guarantee of the accuracy of the level of statistical variation modeled in manufacturing.

The main drawback of purely numerical approaches is the significant amount of effort required to construct the numerical models. The RSM approach of Power (1997) required SPICE model extractions from 100 die. In addition, process changes are made throughout the life of a technology, and the confidence that the effects of these changes can be quickly and accurately reflected in models is significantly greater for physically based models than for purely numerical models.

6. TYPICAL CASE MODELING

All statistical models are generated with respect to a "typical" case model that represents a device with mean values of ***e***. Often, the typical (or "nominal") case model file is derived by extracting models from a "golden" wafer, that is meant to represent typical manufacturing conditions. However, for the reasons enumerated in section 5.2, it is never possible to guarantee that a given wafer and site are exactly typical for every element of ***e***.. So extraction from a golden wafer only provides a near typical model, yet identification of this wafer can entail significant time and expense.

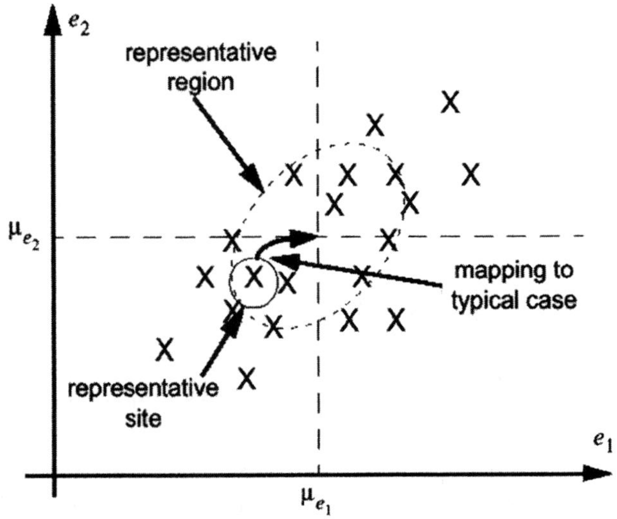

Figure 2. Region for extraction of representative model.

A more accurate and more efficient method to generate a typical case model is as follows. As Figure 2 shows, manufacturing produces wafers and sites with a range of values of ***e***. Based on the process and geometry level models discussed in section 4 it is possible to use nonlinear least squares optimization to adjust ***p*** so that the simulated ***e*** exactly match the mean values of the ***e*** measured. The procedure is detailed in section 9. Therefore models can be extracted from a "representative" wafer and site, and then adjusted to match the mean of . The wafer and site used for extraction of the representative model only need to be screened to make sure they are not too far from typical manufacturing. This requires far less effort than searching for a golden wafer and site, and also guarantees that all ***e*** for the typical case file are at their nominal value, which is not possible with typical case model files extracted from a golden wafer.

Note that extrapolation of MOSFET models to shorter channel lengths than they were extracted from is dangerous, therefore for MOSFET characterization it is desirable to include a device whose channel length is shorter than the allowed minimum for a particular technology, or to extract the representative parameters from a wafer and site with effective channel lengths near the lower limit for the technology. The former approach is preferable.

7. DISTRIBUTIONAL STATISTICAL MODELING

Distributional statistical modeling is required for Monte Carlo type statistical simulation. The goal is therefore to characterize the means and variances of the process parameters $\boldsymbol{p}$. As shown in section 5.3 direct measurement of $\boldsymbol{p}$ does not guarantee accurate modeling of the statistical distributions of $\boldsymbol{e}$, which is the real goal of distributional statistical modeling. This is because the mappings from $\boldsymbol{p}$ to $\boldsymbol{s}$ and from $\boldsymbol{s}$ to $\boldsymbol{e}$ are approximate, therefore forward propagation of variance is inaccurate. Instead, as Figure 3 shows, what is required is some method to be able to calculate the distributions of $\boldsymbol{p}$ from the distributions of $\boldsymbol{e}$.

Consider the δe_i variation in one element e_i of $\boldsymbol{e}$ caused by perturbations δp_k in the elements of $\boldsymbol{p}$, Sensitivity analysis gives

$$\delta e_i = \sum_k \frac{\partial e_i}{\partial p_k} \delta p_k . \tag{7}$$

Propagation of variance gives

$$\sigma^2_{\delta e_i} = \sum_k \left(\frac{\partial e_i}{\partial p_k}\right)^2 \sigma^2_{\delta p_k} \tag{8}$$

which defines a set of linear equations that relate the variances in $\delta\boldsymbol{e}$ to the variances in $\delta\boldsymbol{p}$ The sensitivities in equation (8) are computed by differencing, and the variances in are specified, as the distributions of the PC data. Equation (8) can therefore be directly solved for the variances of the process parameters. A specific example is given below in section 10.

The above technique for distributional statistical characterization is called backward propagation of variance (BPV).

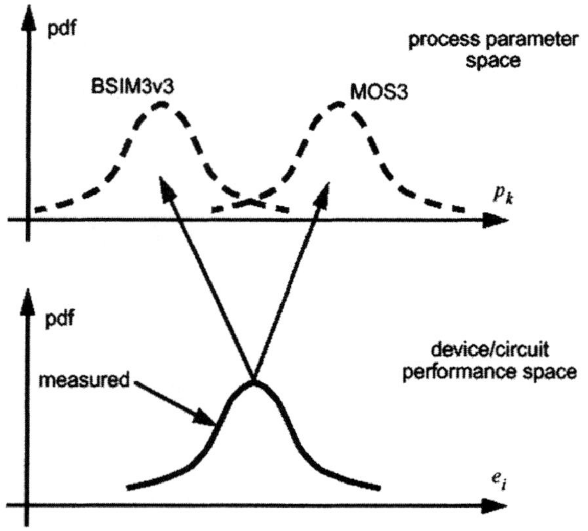

Figure 3. Statistical characterization using backward propagation of variance.

The number of e_i can be greater than the number of p_k, this merely gives an overdetermined set of equations (8) that can be solved by standard techniques.

Several issues arise in solving for $\sigma_{\delta pk}$. First, $\boldsymbol{p}$ must be mathematically observable in $\boldsymbol{e}$. This does not mean that there must be a one to one correspondence between each p_k and one individual e_i. It means that the matrix of squared sensitivities in equation (8) must be mathematically well conditioned, at the very least nonsingular. Consequently, the PC data used to statistically characterize the process parameters for each type of device being modeled must be selected carefully to make the $\boldsymbol{p}$ observable. With a physical understanding of the way the process parameters affect device behavior this is generally easy to do. PC data suitable for MOSFET statistical characterization are presented in section 10, and PC data suitable for BJT statistical characterization are given in McAndrew (1997).

Second, note that not all process parameters must be characterized statistically using BPV. For resistors, it is adequate to directly measure the statistical variations of ρ_s, ΔL_r and ΔW_r. For MOSFETs it is best to directly measure the distribution of T_{ox}, from capacitance data, and then this gives terms that subtract form the left hand side of equation (8), as section 10 shows. So statistical characterization can be a mix of forward and backward propagation of variance.

Third, for most quantities it is natural to think of variations in a relative sense, i.e. in terms of a percentage variation. For these quantities the variations and sensitivities in equation (8) should be normalized. This typically scales the sensitivities to order unity, and therefore improves the numerical stability of the solution of equation (8).

For MOSFET statistical modeling variations in flatband voltage Vfb and lateral geometry Δ_L and Δ_W are the main process parameters that should be modeled in an absolute sense, and are not normalized.

Fourth, the mappings from $\boldsymbol{p}$ to $\boldsymbol{e}$ are close to, but not exactly, linear. This means that the sensitivity estimates from differencing, and consequently the standard deviations $\sigma_{\delta pk}$, depend on the range of perturbations used in $\delta\boldsymbol{p}$. This range is not known *a priori*. To ensure a self consistent solution of equation (8), an initial estimate of each $\sigma_{\delta pk}$ is made, and the sensitivities are calculated using $\pm 3\sigma$ variations in each p_k about its typical value (0 or 1 for elements of $\delta\boldsymbol{p}$ best considered in an absolute or relative sense, respectively). This allows an update in the values of $\sigma\delta p_k$ by solving equation (8), and a corresponding update in the $\pm 3\sigma$ perturbations used to calculate the sensitivities, and a subsequent update of $\sigma_{\delta pk}$. This procedure can be repeated iteratively until the updates are small. Experience has shown that the mappings are relatively linear, and only one update cycle is required.

In certain circumstances the solution of equation (8) can fail miserably. This may seem like a major problem, however it is actually one of the most powerful advantages of the BPV approach, as it allows diagnosis of problems with data that are input to the statistical modeling procedure. No other method for statistical modeling has this useful and desirable property.

Two types of problems can be detected. First, if the matrix of squared sensitivities is poorly conditioned, which becomes evident when solving equation (8), then the selection of the $\boldsymbol{e}$ for statistical modeling is insufficient. In BJTs an increase in the pinched base sheet resistance ρ_{sbe} causes an increase in both the collector current I_C and the intrinsic base resistance R_{BI}. At high current levels the increase in R_{BI} causes a debiasing of the intrinsic base-emitter junction, which lowers I_C . This means that there is a point at which the sensitivity $\partial I_c / \partial \rho_{sbe}$, which appears in equation (8) for BJT statistical modeling (McAndrew, 1997), is zero, which makes the matrix of squared sensitivities singular. This specific BJT modeling problem is solved by reducing the base-emitter bias at which I_C is measured, and a general solution requires analysis of the matrix of squared sensitivities and of $\boldsymbol{e}$

Second, if there are inconsistencies in the input variances of $\boldsymbol{e}$, then solution of equation (8) can yield negative variances for some elements of $\boldsymbol{p}$, which is clearly absurd. For example, variations in I_{dsat} for a wide/long MOSFET are predominantly caused by variations in T_{ox}, with some minor sensitivity to the flatband voltage V_{fb} and effective substrate doping N_b. If the standard deviation of T_{ox} is specified as being greater than the standard deviation of T_{ox} for a wide/long device, then this is physically inconsistent and leads to negative variances when equation (8) is solved. This is easily detected and flagged by the program that characterizes the statistical models. Analysis for BJTs (McAndrew, 1997) shows that

$$\rho^2_{\delta\beta/\beta} < \rho^2_{\delta I_c/I_c} + \rho^2_{\delta I_b/I_b} \tag{9}$$

must hold as a consistency relation between the input variances.

This second problem is not uncommon. It seems to be typical for manufacturing to specify a range for T_{ox} variation that is larger, by a factor of 2 or 3, than the actual control capability of a technology. Using such a specification along with measured manufacturing variations in I_{dsat} and threshold voltage V_{th} is guaranteed to lead to physical inconsistencies in variances of ***e***. It is a deep and powerful property of the above analyses that these inconsistencies can be automatically detected.

8. SPECIFIC CASE STATISTICAL MODELING

When designing an IC it is desirable to know the expected manufacturing variation of every important measure of circuit performance. In general, different points, at a specified probability level in ***p*** space, define the extreme values for different measures of circuit performance. Therefore for accurate case based design a specific case file should be used for every e_i for every circuit.

As Figure 4 shows, the problem can be defined as

$$\max_{\delta \boldsymbol{p}} \delta e_i - \sum_k \frac{\partial e_i}{\partial p_k} \delta p_k \text{ subject to } \delta \boldsymbol{p}^T \boldsymbol{C}^{-1} \delta \boldsymbol{p} - z^2 \tag{10}$$

where $\boldsymbol{C}$ is the variance/covariance matrix of $\delta \boldsymbol{p}$ and z is the sigma level in the manufacturing space of ***p*** at which the maximum of δe_i is being sought. Note that here it is not assumed that the $\delta \boldsymbol{p}$ are uncorrelated. The formulation of equation (10) is the inverse of the statistical design problem presented by Nassif (1994).

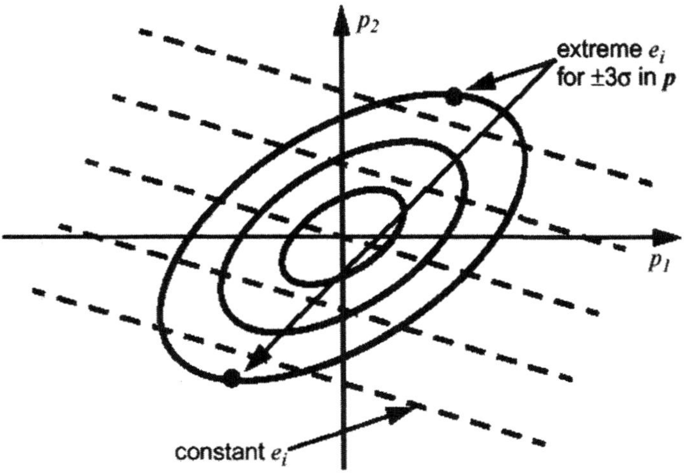

Figure 4. Equiprobability $\boldsymbol{p}$ contours and constant *ei* loci.

$$\frac{\partial L}{\partial \delta \boldsymbol{p}} = \frac{\partial e_i}{\partial \boldsymbol{p}} + 2\lambda \boldsymbol{C}^{-1}\delta \boldsymbol{p} = \boldsymbol{0} \tag{12}$$

$$\frac{\partial L}{\partial \lambda} = \delta \boldsymbol{p}^T \boldsymbol{C}^{-1}\delta \boldsymbol{p} - z^2 = 0 \tag{13}$$

Forming the Lagrangian
for the solution of equation (10). The result is

$$\delta \boldsymbol{p} = \pm z \frac{\boldsymbol{C}(\partial e_i/\partial \boldsymbol{p})}{\sqrt{(\partial e_i/\partial \boldsymbol{p})^T \boldsymbol{C}(\partial e_i/\partial \boldsymbol{p})}}, \tag{14}$$

and for the case of uncorrelated $\boldsymbol{p}$ this reduces to

$$\delta p_j = \pm z \frac{\sigma^2_{\delta p_j}(\partial e_i/\partial p_j)}{\sqrt{\sum_k \sigma^2_{\delta p_k}(\partial e_i/\partial p_k)^2}}. \tag{15}$$

Intuitively equation (15) makes sense. If e_i depends on only one element p_k of $\boldsymbol{p}$, so that $\partial e_i / \partial p_j$ is zero for all except $j = k$, then the solution is $\delta p_j = 0$ for $j \neq k$ and $\delta p_k = \pm z\sigma_{\delta pk}$, as expected. Although formulated as the inverse of the statistical design problem of Nassif (1994), the result (14) is identical, again as could be intuitively expected.

To solve for specific case statistical models, the sensitivities and variances in equation (15) are determined as in section 7, and then equation (15) is solved for the $\delta \boldsymbol{p}$ that define the case files. Comparison of computed and linearly predicted values for e_i at the extreme cases has verified that the assumed linear relationship between $\boldsymbol{e}$ and $\boldsymbol{p}$ is reasonable (McAndrew, 1997).

In practice, specific case statistical models are not often used, because they need to be calculated uniquely for individual circuits. Generic case models (see section 9) are preferred. However, the procedure detailed here allows simple generation of specific case models when they are needed. In particular, because the procedure explicitly uses the sensitivities $\partial e_i / \partial p_j$ it avoids the problem described in section 5.3 of perturbing p_k when it does not affect e_i. In addition, the procedure is very efficient because it requires only enough simulations to allow the sensitivities to be calculated. For n process parameters this can be as few as n+1 simulations, depending on the differencing method used to calculate $\partial e_i / \partial p_j$, and is independent of the number of measures of circuit performance $\boldsymbol{e}$ for which specific case files are to be generated.

Note that the assumption of linearity between $\boldsymbol{p}$ and $\boldsymbol{e}$ is good for $\boldsymbol{e}$ that are measures of device electrical performance. Some circuits have performance measures that do not behave as nicely, and these require distributional statistical simulation for accurate statistical design (Rencher, 1989).

9. GENERIC CASE STATISTICAL MODELING

Generic case statistical models are intended to model specified variation levels, typically $\pm 3\sigma$, in each element of $\boldsymbol{e}$. As noted in section 2, these case files do not guarantee $\pm 3\sigma$ modeling of every measure of circuit performance for every circuit. However, they can provide a reasonable and efficient means to uncover undesired circuit sensitivities, and proximity to regions of catastrophic failure. And they can work well for some types of circuits, e.g. digital CMOS.

The method to determine generic case files is simple. Using nonlinear least squares optimization, the process parameters $\boldsymbol{p}$ are adjusted until the $\boldsymbol{e}$ are at the desired $\pm 3\sigma$ limit for a particular case. In some cases, the need $\boldsymbol{e}$ to be selected differently for this purpose than for generating distributional statistical models. For example, for BJTs , and I_c, I_b and $\beta = I_c / I_b$ and allow distributional statistical characterization, but are clearly dependent when it comes to characterizing generic case models (McAndrew, 1997).

Note that this procedure can generate somewhat unphysical models in some cases. For example, for MOSFETs criteria for generating generic case files include ±3σ values for I_{dsat} for both wide/long and wide/short devices. Given that the sensitivities of I_{dsat} to variations in Δ_L and T_{ox} differ between long and short devices, it is apparent that manufacturing that leads to ±3σ variations in I_{dsat} for a wide/short device will in general not be consistent with ±3σ variations in I_{dsat} for a wide/long device. However, the resultant generic case files are absolutely guaranteed to span the ±3σ in all $\boldsymbol{e}$, as is desired. The specific case files defined in section 8 are guaranteed to be physically consistent for all $\boldsymbol{e}$, but they only provide ±3σ variation modeling for one e_i at a time.

10. SPECIFIC MOSFET EXAMPLE

The key process parameters $\boldsymbol{p}$ that control the variation of MOSFET electrical behavior are the oxide thickness T_{ox}, the flatband voltage V_{fb}, the channel length reduction Δ_L, the channel width variation Δ_W ($W_{eff} = W_g + \Delta_W$), the effective substrate doping N_b, the low field mobility μ_0, and a parameter V_{tl} that models the reduction in threshold voltage V_{th0} with decreasing channel length, and depends on junction depth, profile shape, etc. For most MOSFET models there is a direct, one to one, correspondence between some of these process parameters and the model parameters. For some model parameters $\boldsymbol{s}$ mappings may need to be defined from the process parameters $\boldsymbol{p}$ to $\boldsymbol{s}$, e.g.

$$V_{th} = V_{fb} + \phi_B + \frac{T_{ox}\sqrt{2q\varepsilon_{Si}N_b}}{\varepsilon_{ox}}\sqrt{\phi_B}, \quad \phi_B = 2\frac{kT}{q}\ln\left(\frac{N_b}{n_i}\right) \tag{16}$$

and N_b needs to be mapped into surface and bulk doping levels if the SPICE model on top of which the statistical model is being built is based on a two level channel doping profile model. V_{tl} must be mapped into the parameters that control V_{th0} over channel length for a specific model.

Here variations in T_{ox} and Δ_W will be used directly, by forward propagation of variances. (It is also possible to characterize the statistics of Δ_W variation by including as a device electrical performance a direct measure of drain current for a narrow/long MOSFET). Statistical modeling of V_{fb}, Δ_L, N_b, μ_0 and V_{tl} will be done based on backward propagation of variance.

Consideration of how each of these parameters influences device behavior makes it apparent that they observable, in the sense defined in section 7, in the device electrical performances V_{tr}, the threshold voltage of a wide/long (reference) device, I_{dr}, the saturated drain current of a wide/long device, V_{ts} and I_{ds}, the threshold voltage and saturated drain current of wide/short device, and the body effect D_{tr}, the difference in threshold voltage for a wide/long device between V_{sb} at V_{DD} (or its maximum allowable value) and zero. There is not a one to one correspondence between ***p*** and ***e*** However, as Figure 5 shows, the effects of variations in ***p*** can be unscrambled from variations in ***e***. The solid lines represent strong dependencies, the dashed lines are weak dependencies.

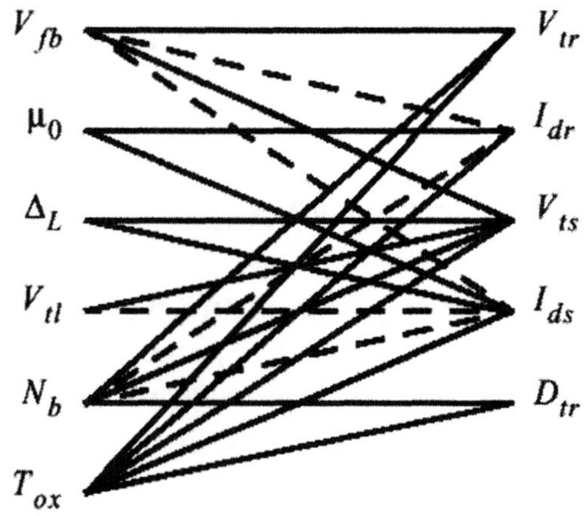

Figure 5. Dependence of MOSFET electrical performances on process parameters.

The sensitivity equations (7) for this example are

$$\begin{bmatrix} \delta V_{tr} \\ \dfrac{\delta I_{dr}}{I_{dr}} \\ \delta V_{ts} \\ \dfrac{\delta I_{ds}}{I_{ds}} \\ \delta D_{tr} \end{bmatrix} = \begin{bmatrix} \dfrac{\partial V_{tr}}{\partial V_{fb}} & \mu_0 \dfrac{\partial V_{tr}}{\partial \mu_0} & \dfrac{\partial V_{tr}}{\partial \Delta_L} & \dfrac{\partial V_{tr}}{\partial V_{tl}} & N_b \dfrac{\partial V_{tr}}{\partial N_b} & T_{ox} \dfrac{\partial V_{tr}}{\partial T_{ox}} \\ \dfrac{1}{I_{dr}} \dfrac{\partial I_{dr}}{\partial V_{fb}} & \dfrac{\mu_0}{I_{dr}} \dfrac{\partial I_{dr}}{\partial \mu_0} & \dfrac{1}{I_{dr}} \dfrac{\partial I_{dr}}{\partial \Delta_L} & \dfrac{1}{I_{dr}} \dfrac{\partial I_{dr}}{\partial V_{tl}} & \dfrac{N_b}{I_{dr}} \dfrac{\partial I_{dr}}{\partial N_b} & \dfrac{T_{ox}}{I_{dr}} \dfrac{\partial I_{dr}}{\partial T_{ox}} \\ \dfrac{\partial V_{ts}}{\partial V_{fb}} & \mu_0 \dfrac{\partial V_{ts}}{\partial \mu_0} & \dfrac{\partial V_{ts}}{\partial \Delta_L} & \dfrac{\partial V_{ts}}{\partial V_{tl}} & N_b \dfrac{\partial V_{ts}}{\partial N_b} & T_{ox} \dfrac{\partial V_{ts}}{\partial T_{ox}} \\ \dfrac{1}{I_{ds}} \dfrac{\partial I_{ds}}{\partial V_{fb}} & \dfrac{\mu_0}{I_{ds}} \dfrac{\partial I_{ds}}{\partial \mu_0} & \dfrac{1}{I_{ds}} \dfrac{\partial I_{ds}}{\partial \Delta_L} & \dfrac{1}{I_{ds}} \dfrac{\partial I_{ds}}{\partial V_{tl}} & \dfrac{N_b}{I_{ds}} \dfrac{\partial I_{ds}}{\partial N_b} & \dfrac{T_{ox}}{I_{ds}} \dfrac{\partial I_{ds}}{\partial T_{ox}} \\ \dfrac{\partial D_{tr}}{\partial V_{fb}} & \mu_0 \dfrac{\partial D_{tr}}{\partial \mu_0} & \dfrac{\partial D_{tr}}{\partial \Delta_L} & \dfrac{\partial D_{tr}}{\partial V_{tl}} & N_b \dfrac{\partial D_{tr}}{\partial N_b} & T_{ox} \dfrac{\partial D_{tr}}{\partial T_{ox}} \end{bmatrix} \begin{bmatrix} \delta V_{fb} \\ \dfrac{\delta \mu_0}{\mu_0} \\ \delta \Delta_L \\ \delta V_{tl} \\ \dfrac{\delta N_b}{N_b} \\ \dfrac{\delta T_{ox}}{T_{ox}} \end{bmatrix} \qquad (17)$$

where the appropriate normalization has been used, and because all ***e*** are for wide devices there is negligible dependence on Δ_W. The equations solved for BPV are

$$\begin{bmatrix} \sigma^2_{\delta V_{tr}} - \left(T_{ox}\frac{\partial V_{tr}}{\partial T_{ox}}\right)^2 \sigma^2_{\delta T_{ox}/T_{ox}} \\ \sigma^2_{\delta I_{dr}/I_{dr}} - \left(\frac{T_{ox}}{I_{dr}}\frac{\partial I_{dr}}{\partial T_{ox}}\right)^2 \sigma^2_{\delta T_{ox}/T_{ox}} \\ \sigma^2_{\delta V_{ts}} - \left(T_{ox}\frac{\partial V_{ts}}{\partial T_{ox}}\right)^2 \sigma^2_{\delta T_{ox}/T_{ox}} \\ \sigma^2_{\delta I_{ds}/I_{ds}} - \left(\frac{T_{ox}}{I_{ds}}\frac{\partial I_{ds}}{\partial T_{ox}}\right)^2 \sigma^2_{\delta T_{ox}/T_{ox}} \\ \sigma^2_{\delta D_{tr}} - \left(T_{ox}\frac{\partial D_{tr}}{\partial T_{ox}}\right)^2 \sigma^2_{\delta T_{ox}/T_{ox}} \end{bmatrix} = \tag{18}$$

$$\begin{bmatrix} \left(\frac{\partial V_{tr}}{\partial V_{fb}}\right)^2 & \left(\mu_0\frac{\partial V_{tr}}{\partial \mu_0}\right)^2 & \left(\frac{\partial V_{tr}}{\partial \Delta_L}\right)^2 & \left(\frac{\partial V_{tr}}{\partial V_{tl}}\right)^2 & \left(N_b\frac{\partial V_{tr}}{\partial N_b}\right)^2 \\ \left(\frac{1}{I_{dr}}\frac{\partial I_{dr}}{\partial V_{fb}}\right)^2 & \left(\frac{\mu_0}{I_{dr}}\frac{\partial I_{dr}}{\partial \mu_0}\right)^2 & \left(\frac{1}{I_{dr}}\frac{\partial I_{dr}}{\partial \Delta_L}\right)^2 & \left(\frac{1}{I_{dr}}\frac{\partial I_{dr}}{\partial V_{tl}}\right)^2 & \left(\frac{N_b}{I_{dr}}\frac{\partial I_{dr}}{\partial N_b}\right)^2 \\ \left(\frac{\partial V_{ts}}{\partial V_{fb}}\right)^2 & \left(\mu_0\frac{\partial V_{ts}}{\partial \mu_0}\right)^2 & \left(\frac{\partial V_{ts}}{\partial \Delta_L}\right)^2 & \left(\frac{\partial V_{ts}}{\partial V_{tl}}\right)^2 & \left(N_b\frac{\partial V_{ts}}{\partial N_b}\right)^2 \\ \left(\frac{1}{I_{ds}}\frac{\partial I_{ds}}{\partial V_{fb}}\right)^2 & \left(\frac{\mu_0}{I_{ds}}\frac{\partial I_{ds}}{\partial \mu_0}\right)^2 & \left(\frac{1}{I_{ds}}\frac{\partial I_{ds}}{\partial \Delta_L}\right)^2 & \left(\frac{1}{I_{ds}}\frac{\partial I_{ds}}{\partial V_{tl}}\right)^2 & \left(\frac{N_b}{I_{ds}}\frac{\partial I_{ds}}{\partial N_b}\right)^2 \\ \left(\frac{\partial D_{tr}}{\partial V_{fb}}\right)^2 & \left(\mu_0\frac{\partial D_{tr}}{\partial \mu_0}\right)^2 & \left(\frac{\partial D_{tr}}{\partial \Delta_L}\right)^2 & \left(\frac{\partial D_{tr}}{\partial V_{tl}}\right)^2 & \left(N_b\frac{\partial D_{tr}}{\partial N_b}\right)^2 \end{bmatrix} \begin{bmatrix} \sigma^2_{\delta V_{fb}} \\ \sigma^2_{\delta \mu_0/\mu_0} \\ \sigma^2_{\delta \Delta_L} \\ \sigma^2_{\delta V_{tl}} \\ \sigma^2_{\delta N_b/N_b} \end{bmatrix}$$

where the variation caused by T_{ox} is subtracted from the left hand side. Because of both the normalization and the weak dependence of some e_i on p_k many of the entries of the matrix of squared sensitivities in equation (18) are close to either 1 or 0.

The BPV procedure for generating distributional models for MOSFETs and BJTs has been in use at Motorola since 1995. For one particular technology NMOS and PMOS statistical models were characterized from manufacturing PC data, and then a 1000 sample Monte Carlo simulation was done to verify the results of the statistical models (the manufacturing data did not include D_{tr}). The statistical modeling was based on both the SSIM model (Veeraraghavan, 1990) and the MOS3 or SPICE level 3 model (Vladimirescu, 1980). Table 2 shows the accuracy of the models.

Table 2. Comparison of measured and modeled MOSFET statistics

Parameter	fab mean	SSIM mean	MOS3 mean	Fab Sigma	SSIM sigma	MOS3 sigma
NMOS V_{tr} (V)	0.950	0.950	0.953	0.0300	0.0287	0.0286
NMOS I_{dr} (mA)	1.010	1.009	1.009	0.0400	0.0399	0.0406
NMOS V_{ts} (V)	0.350	0.350	0.350	0.0158	0.0153	0.0153
NMOS I_{ds} (mA)	19.000	19.069	19.032	1.3300	1.3291	1.3307
PMOS V_{tr} (V)	1.050	1.050	1.049	0.0600	0.0616	0.0620
PMOS I_{dr} (mA)	0.150	0.151	0.151	0.0075	0.0074	0.0075
PMOS V_{ts} (V)	1.000	0.998	1.005	0.0800	0.0794	0.0707
PMOS I_{ds} (mA)	11.500	11.603	11.594	1.1500	1.1154	1.1010

Figures 6 through 9 show the Monte Carlo simulation data, along with the manufacturing ±3σ limits as dashed lines. T_{ox} and ΔL were correlated between the NMOS and PMOS devices. The accuracy of the modeling is clear.

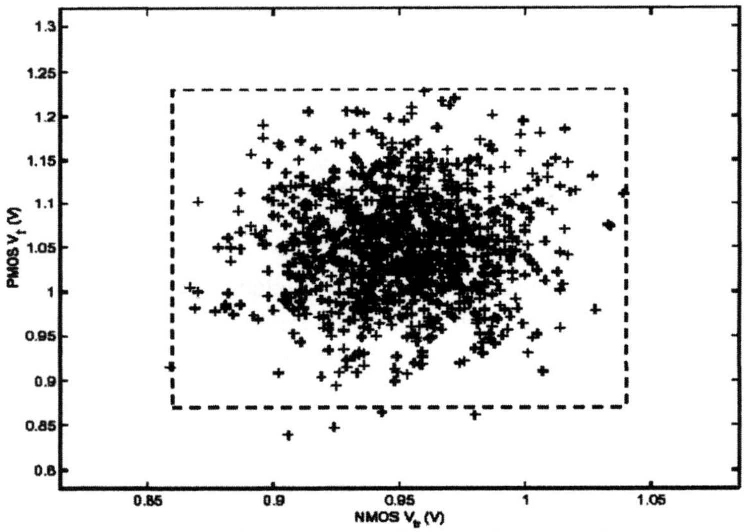

Figure 6. Wide/long PMOS and NMOS threshold voltage.

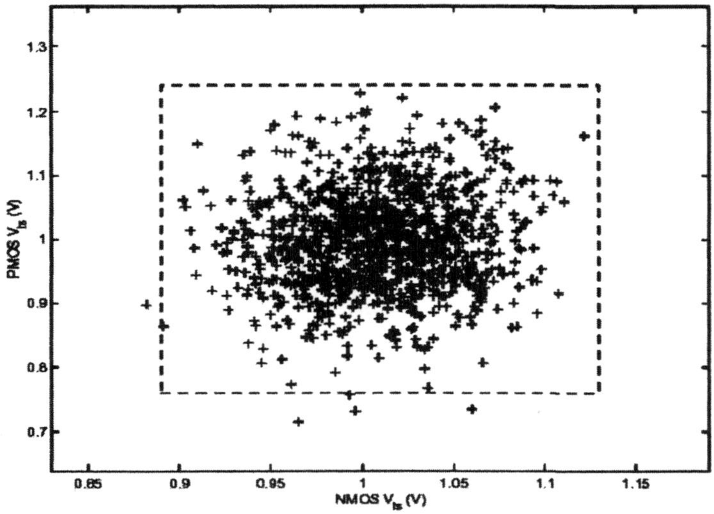

Figure 7. Wide/short PMOS and NMOS threshold voltage.

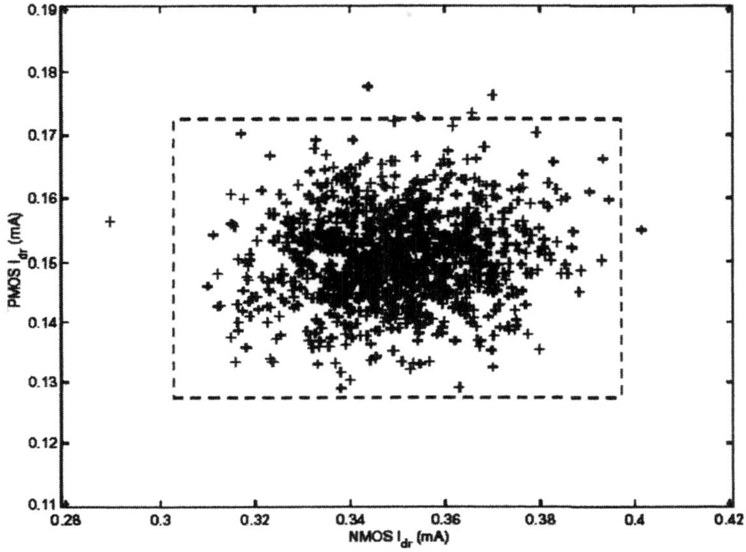

Figure 8. Wide/long PMOS and NMOS saturated drain current.

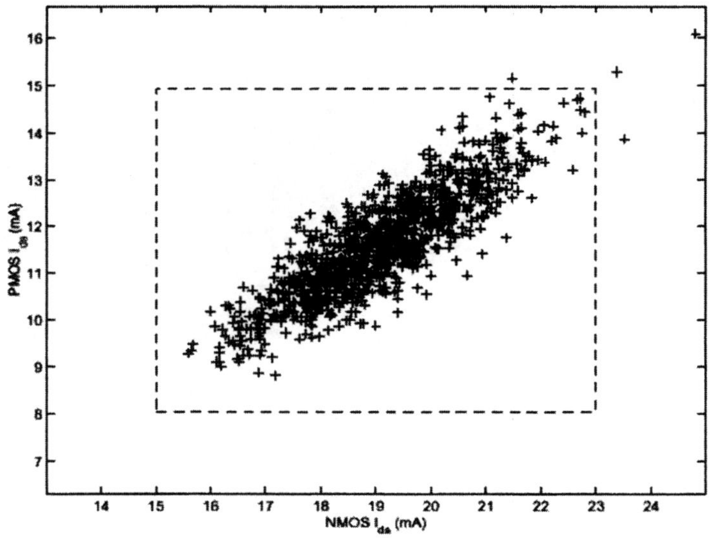

Figure 9. Wide/short PMOS and NMOS saturated drain current.

The generic case files were determined by optimizing the process parameters so that ±3σ limits in the manufacturing data were modeled exactly. There was no difference between simulations from the resultant models and the manufacturing limits in Figures 6 through 9. Figures 10 through 13 show simulation results from the generic case files. Figures 10 and 11 show I_d (V_{gs}) curves for wide/long and wide/ short devices, respectively, at V_{ds} = 0.1 and V_{sb} = 0, and Figures 12 and 13 show I_d (V_{ds}) curves for wide/long and wide/short devices, respectively, at V_{gs} = 5.0 and V_{sb} = 0. Clearly, transconductance g_m and output conductance g_o variations also track with the case files.

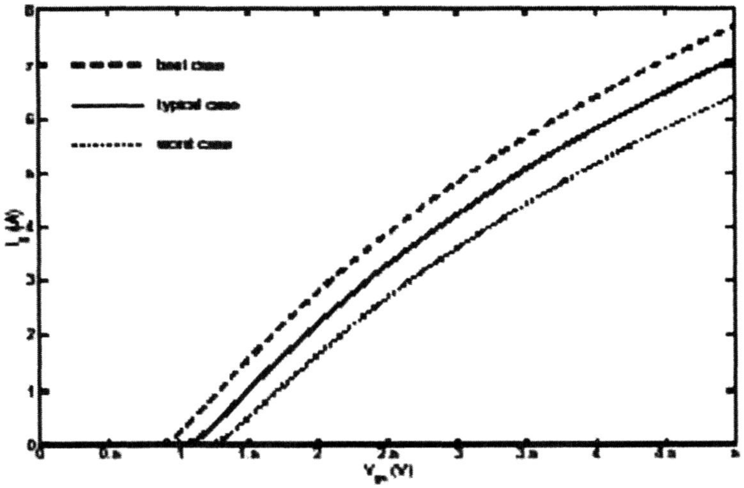

Figure 10. Wide/long PMOS threshold characteristics.

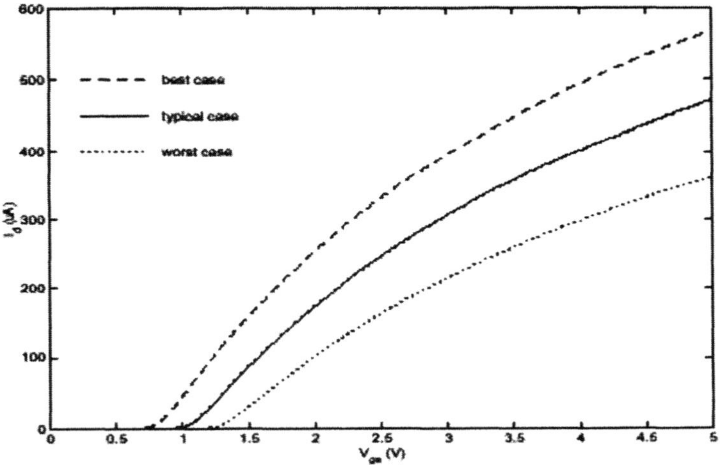

Figure 11. Wide/short PMOS threshold characteristcs

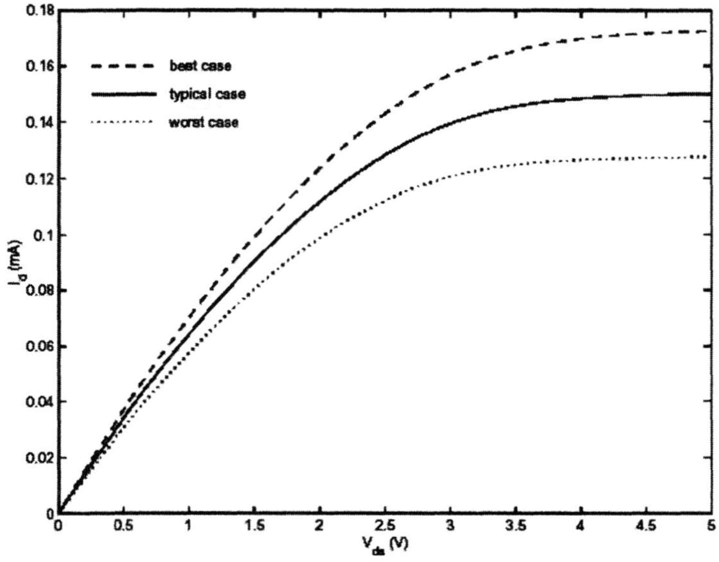

Figure 12. Wide/long PMOS output characteristics

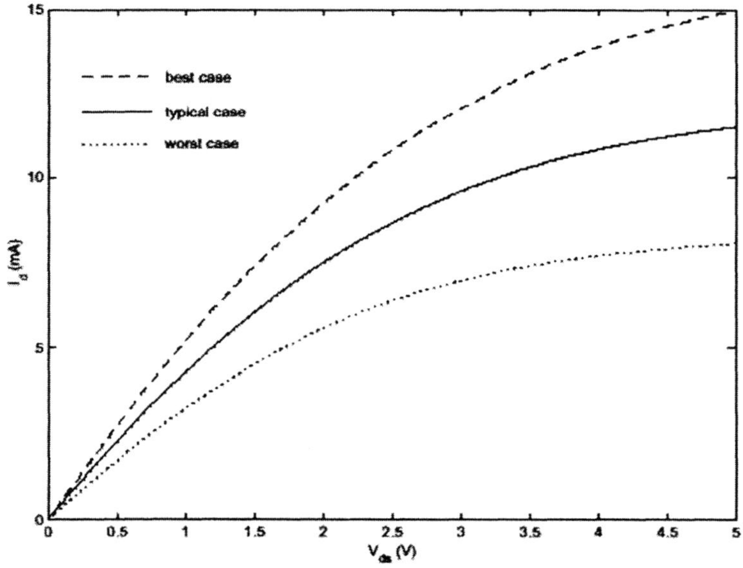

Figure 13. *Wide/short PMOS output characteristics*

11. CONCLUSIONS

Accurate and efficient statistical modeling for circuit simulation should be based on process and geometry level modeling. Different types of statistical models have been explicitly identified, and techniques based on physical process and geometry dependent models, sensitivity analysis, and backward propagation of variance have been presented to allow efficient and accurate characterization of each type of statistical model. These techniques place statistical modeling in a sound theoretical framework, and provide a method to determine if statistical information being input to a statistical characterization procedure is consistent. This has not been possible with previous *ad hoc*, empirical, or strictly numerical approaches.

It has been explicitly recognized that the mappings from process parameter to SPICE model parameters are only approximate, and that the SPICE model representations of real device behavior are also only approximate. Therefore it is inefficient to expend significant effort to try to directly model process parameter variations, or to construct accurate numerical versions of the $\boldsymbol{p}$ to $\boldsymbol{s}$ mappings as is done by PCA or RSM techniques. The approaches presented here in essence "take up the slack" in the errors in the deterministic $\boldsymbol{e}(\boldsymbol{s}(\boldsymbol{p}))$models, and efficiently give statistical models that are guaranteed to model $\boldsymbol{e}$ accurately, which is the real goal of statistical modeling for circuit simulation.

Many analog IC design techniques are based on mismatch between components, because component mismatch is much smaller in IC manufacturing technologies than is the variation in the absolute value of device parametrics. Mismatch analysis and modeling follow exactly the same procedures presented here for statistical modeling, with the variances being interpreted as between devices rather than as being for large scale manufacturing variations.

12. REFERENCES

Chen, J. C., Hu, C., Liu, Z., and Ko, P. K. (1995) Realistic worst-case SPICE file extraction using BSIM3. *Proc. IEEE CICC*, 375-8.

Cox, P., Yang, P., Mahant-Shetti, S. S., and Chatterjee, P. (1985) Statistical modeling for efficient parametric yield estimation of MOS VLSI circuits. *IEEE Trans. Electron Dev.*, **ED-32**, 471-8.

Davis, W. F. and Ida, R. T. (1989) Statistical IC simulation based on independent wafer extracted process parameters and experimental designs. *Proc. IEEE BCTM*, 262-5.

McAndrew, C. C., Bates, J., Ida, R. T., and Drennan, P. (1997) Efficient statistical BJT modeling, why • is more than Ic/Ib. *Proc. IEEE BCTM*.

Nagel, L. W. (1975) SPICE2: A computer program to simulate semiconductor circuits. Memo. no. ERL-520, Electronics Research Laboratory, University of California, Berkeley.

Nassif, S. R. (1994) Statistical worst-case analysis for integrated circuits, in *Statistical Approach to VLSI* (eds. S. W. Director, W. Maly, and A. Strowjas), North-Holland, 233-53.

Ogrodzki J., Opalski, L., and Styblinski, M. (1980) Acceptability regions for a class of linear networks. *Proc. IEEE ISCAS*, 187-90.

Papoulis, A. (1991) Probability, random variables, and stochastic processes, 3rd. edition. McGraw-Hill, New York.

Power, J. A., Donellan, B., Mathewson, A., and Lane, W. A. (1994) Relating statistical MOSFET model parameter variabilities to IC manufacturing process fluctuations enabling realistic worst-case design. *IEEE Trans. Semicond. Manufact.*, **7**, 306-18.

Power, J. A., Kelly, S., Griffith, E., Doyle, D., and O'Neill, M. (1997) Statistical modeling for 0.6μm BiCMOS technology. *Proc. IEEE BCTM.*

Prendergast, E. J. and Lloyd, P. (1985) A highly automated integrated modeling system; MECCA. *Proc. IEEE CICC.*

Rencher, M. and Salamina, N. (1989) Statistical bipolar circuit design using MSTAT. *Proc. IEEE ICCAD.*

Veeraraghavan, S. (1990) SSIM: a new charge-based MOSFET model. *MCNC Circuit Simulation Workshop.*

Vladimirescu, A. and Liu, S. (1980) The simulation of MOS integrated circuits using SPICE2. Memo. no. ERL-M80/7, Electronics Research Laboratory, University of California, Berkeley.

13. BIOGRAPHY

Colin McAndrew received the Ph.D. and M.A.Sc. degrees in systems design engineering from the University of Waterloo, Waterloo, Ont., Canada, in 1984 and 1982, and the B.E. degree in electrical engineering from Monash University, Melbourne, Vic., Australia, in 1978. Since 1995 he has been the Manager of the Statistical Modeling and Characterization Laboratory at Motorola, Tempe AZ. From 1987 to 1995 he was a Member of Technical Staff at AT&T Bell Laboratories, Allentown PA. From 1984 to 1987 and 1978 to 1980 he was an engineer at the Herman Research Laboratory of the State Electricity Commission of Victoria.

Chapter 5

Retargetable Application-driven Analog-digital Block Design

J. E. Franca
Instituto Superior Técnico; Av. Rovisco Pais, 1, 1096 Lisboa Codex, Portugal; Tel.: (+351-1) 841.76.77 Fax: (+351-1) 841.76.75; E-mail: franca@ecsm4.ist.utl.pt

Key words: Retargetable design, design for re-use, mixed-signal design, analog-digital blocks

Abstract: In a world subject to a fast pace of technology evolution towards system level integration, drastic increase of design productivity is needed to reduce product development cycles and improve timely availability to the market. This can be achieved through the type of retargetable analog-digital blocks discussed in this chapter, whereby various functional building blocks, such as data conversion, amplification, and filtering, among other functions, are efficiently combined to allow easy re-usability for different technology environments and application requirements. The four main ingredients of such a methodology include optimized system-level partitioning for efficient handling of inter-function design constraints, technology adaptation capability ensured by component design methodologies which reduce technology dependency, effective use of silicon area ensured by maximum layout regularity, optimized component aspect ratios and careful floor planning and, finally, reduced routing density and routing area ensured by detailed analysis of route path characteristics and use of simplifying routing techniques. Examples of practical industry designs are given where such methodology has been efficiently employed to achieve significant improvements in product development cycles.

1. INTRODUCTION

World-wide semiconductor market trends indicate a rapid increase of chips containing both analog and digital functionality, obtained at the expense of chips which contain purely analog functionality and also of chips which contain purely digital functionality. It is generally accepted that such increased functionality is playing a bigger roll in the way integrated circuits are designed, and thereby making designers more concerned with integrated system solutions combining analog and digital functions and signals. In order to provide such increased functionality and combined use of analog and digital signals, it is necessary to develop the electronic circuitry that provides the appropriate analog-digital interface. Thus, in the not too distant future, the integration of complete systems on a chip will be achieved by assembling a variety of high functionality blocks, from powerful CPU and DSP cores to complex analog-digital blocks, as depicted in Figure 1. Such analog-digital blocks correspond to complete sub-systems which embed all the required functionality to interface an analog signal to a digital one and vice versa, including data conversion, filtering and amplification, among other functions. The design of such analog-digital blocks is clearly dependent on the application for which the interfacing function is envisaged.

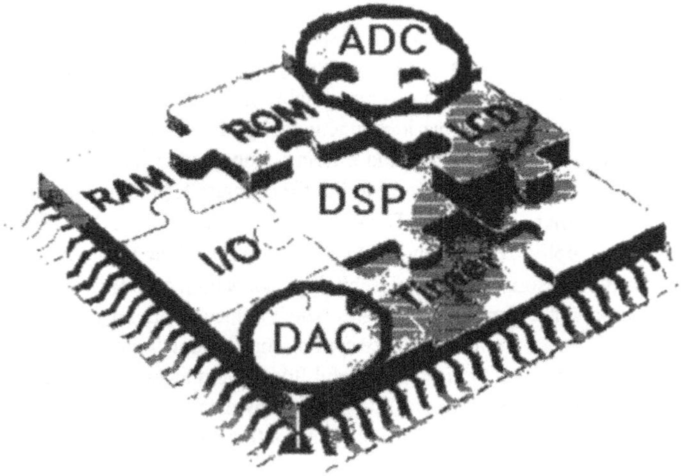

Figure 1. System-level integration in mixed-signal VLSI chips will be achieved using high-functionality analog-digital blocks together with high-density digital cores.

In order to satisfy the need to reduce product development cycles and improve timely availability to the market in a world subject to a fast pace of technology evolution towards system level integration it is inevitably required that a significant increase of productivity is achieved in the design of complex mixed-signal interface functions. This can be provided by retargetable analog-digital blocks that allow easy re-usability for different technology environments and application requirements. This chapter describes the methodology for designing such retargetable application-driven analog-digital blocks, discusses its main ingredients and requirements of the supporting computer-based tools, and illustrates its application in practical industry designs.

2. ANALOG-DIGITAL INTERFACE REQUIREMENTS

Analog-digital blocks found in complex mixed-signal VLSI are needed to interface the digital processing engines at their heart to the physical sources of analog signals. This can be the case of input and output in audio applications, sensors and actuators in micro systems, reading/writing channels for storage applications - disk or tape, and even complete transceiver interfaces for radio transmission. To clarify such requirements, it is worth to consider the two typical interface functions discussed next.

The first is the case of a codec architecture based on delta-sigma technology and which is illustrated by the block diagram shown in Figure 2. The input analog signal is processed through a microphone amplifier and a delta-sigma modulator where such analog signal is digitized in the form of a high-speed bit stream. This is followed by a digital processing unit encompassing both the high-frequency noise filtering and baseband channel shaping. In the complementary interface processing chain, from digital to analog, the block contains a new digital signal processing unit performing baseband channel filtering and pre-distortion, interpolation and truncation, and which is followed by a high-linearity low-resolution digital-to-analog (D/A) converter, an anti-imaging filter, a programmable gain amplifier and a power amplifier driving the output analog port. This shows the high-functionality nature of the block whose implementation requires not only a variety of conversion and signal processing functions, both analog and digital, but also such circuit components as operational amplifiers, comparators and analog switches.

The second example of a mixed-signal interface block, illustrated in Figure 3, provides two fully-differential channels in quadrature phase (90° phase shift) between the input digital ports and the output analog ports. Each channel is formed by a digital processing unit for signal shaping and interpolation, a digital-to-analog conversion function employing current-based circuit techniques and then a continuous-time filter which also provides the embedded current to voltage conversion. An additional calibration unit is employed to adjust the unavoidable offsets and mismatches between the both channels. In order to control the functionality and calibration of the complete block, a control unit is also included. Again in this example we can notice the true multi-functional nature of a block design comprising digital signal processing, digital-to-analog conversion, analog continuous-time filtering and even functions of error correction a nd calibration.

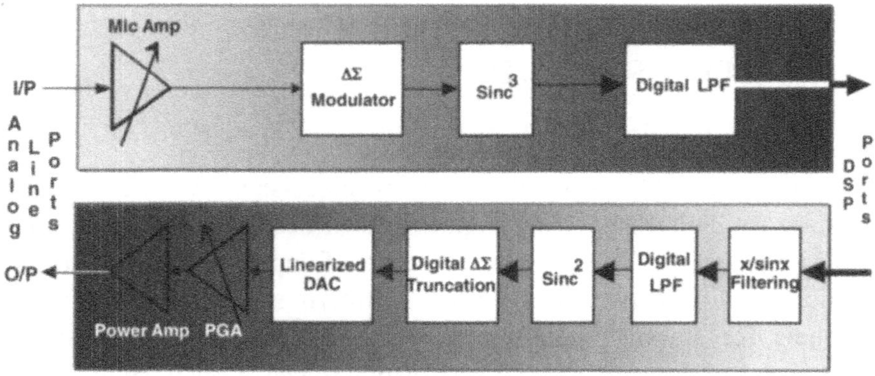

Figure 2. Block diagram of a codec architecture based on delta-sigma technology and which provides full interfacing functionality between two analog ports and two digital ports.

The above examples indicate that mixed-signal interface blocks are very demanding in terms of the knowledge that is needed for their development. Firstly, it is important to have knowledge about the system application and relevant signals, without which it will not be possible to define the block functionality and even its electrical requirements. Once the system is known, it is also important to have a knowledge about the circuits for signal processing and conversion that can be used inside those blocks.

These can include continuous-time and sampled-data filters - both analog and digital, signal conditioning circuits, again both analog and digital, and also analog-to-digital and digital-to-analog conversion functions. Thirdly, it is also important to have the knowledge that allows the development of the circuit components for integrated circuit realization, such as operational amplifiers, comparators, voltage references and current references. Finally, it is important to know how to provide the physical implementation of all those components in a fully verified database that is submitted for fabrication.

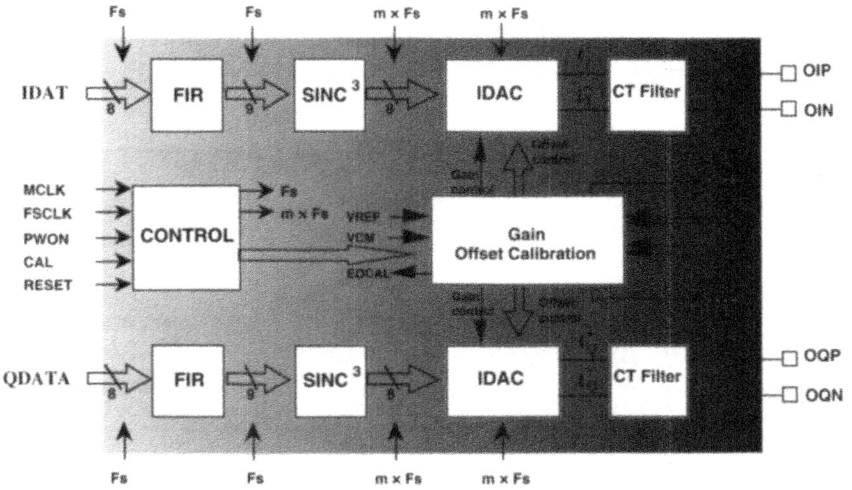

Figure 3. Block diagram of a mixed-signal block providing full interfacing functionality between a quadrature digital port and a quadrature analog port.

3. DESIGN FLOW AND CAD SUPPORT

3.1 Functional hierarchy

An analog-digital block, such as in the previous examples, can be divided into different levels of hierarchy according to the functionality of different elements. This is described in the illustration of Figure 4 for a typical 4-level hierarchy. At the top, there is the system-level where the block functionality is defined. This is composed of different functional blocks whose design is carried-out at a lower level of the hierarchy. At this functional-level there are, for example, conversion functions, filtering functions, both analog and digital, amplification functions, and many others. The design of each one of these functions requires component cells that are defined in a third level of the hierarchy where we can find, for example, operational amplifiers, comparators, reference voltages and current references. Finally, the fourth level of the hierarchy corresponds to the basic elements that are needed to realize circuit components. The basic elements for the realization of mixed-signal integrated circuits are the MOS transistors, resistors, capacitors and maybe even bipolar transistors, if properly supported by the technology.

The integrated circuit design flow and layout implementation crosses the above four levels of the hierarchy. It starts by looking at the system specifications and ends up by having an electric schematic formed by the basic elements. From the electric schematics representing the block the design of the layout that allows the physical implementation in integrated circuit form is carried-out from the bottom to the top of the hierarchy.

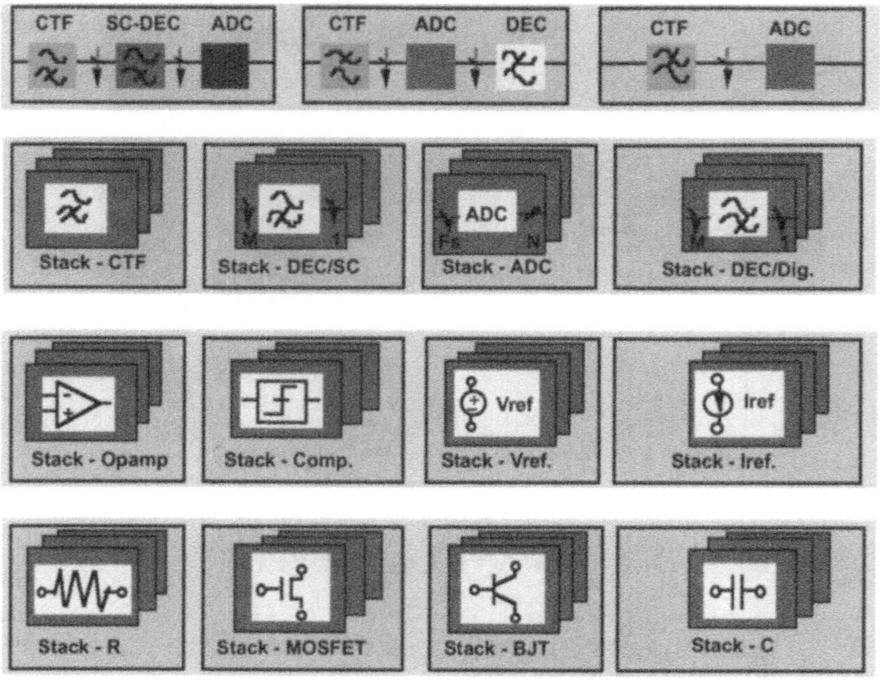

Figure 4. Typical 4-level hierarchy in an analog-digital block.

It starts by laying out the basic elements; then, it networks those elements to form circuit component cells; then, component cells are assembled in a given architecture to form a functional block; finally, various functional blocks are interconnected to form the system. Several tools are currently available to support those flows of information during a complete design cycle of a block: top-down flow for electric design and bottom-up flow for layout design. Their basic characteristics and fundamental limitations for designing analog-digital blocks are discussed next.

3.2 Top-down flow for electrical design

A hierarchical top-design design methodology is typically formed by a synthesis routine encompassing the steps illustrated in Figure 5: (i) architecture selection, (ii) behavioral modeling and simulation, (iii) specification mapping to the lower level of design. The purpose of the CAD tools supporting such a methodology is to speed up the design iterations, not only through powerful computation routines that can help designers explore vast design spaces but also by allowing a reduction of such iterations through improved modeling precision of the designed functions and constituting primitives.

3.2.1 System-level design

The system level design corresponds to the definition of an architecture formed by the interconnection of various functional blocks. Different types of architectures maybe functionally equivalent but lead to very different specifications of the constituting functional blocks, specially from the viewpoint of their integrated circuit realization. It is therefore very important to define the partitioning of the system not only to meet a given set of electrical and functional specifications but also to meet additional practical industrial requirements concerning area consumption and power dissipation. Often, it is also important to define system partitioning in such a way that technology portability can be ensured with minimum additional design intervention.

The CAD tools that are used to support the design at this level are based on behavior modeling and simulation techniques describing interactions between the parameters of the functional building blocks. In recent years, the MatLab/Simulink platform has become a de facto standard to support this level of design, although it is also possible to employ hardware description languages and even write dedicated C-based codes.

3.2.2 Function-level design

Besides the architecture, the system level design also defines the set of specifications associated with each one of the constituting functional blocks. The design of each one of these functions is usually carried-out separately by different designers according to their specific areas of expertise.

This corresponds to the definition of a functional block architecture formed by the interconnection of circuit components, as well as the resulting specifications for each circuit component. Such specifications are traded-off in order to obtain the best solution for integrated circuit implementation. Hence, as before, although different architectures may be functionally equivalent they almost certainly lead to very different practical characteristics regarding area consumption and power dissipation. The verification of this level of design can also be performed by behavioral modeling and simulation and, in some special cases, by full electrical simulation.

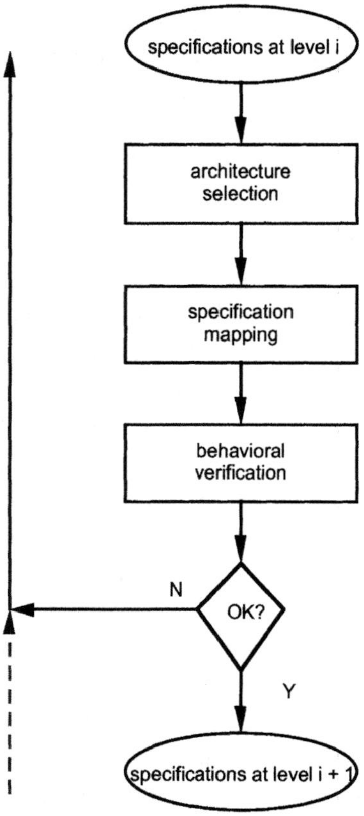

Figure 5. Managing the flow of information between two consecutive levels of an hierarchical design.

Several types of CAD tools have been proposed in the past for the automatic design of functional building blocks, particularly data converters and filters. One type of such CAD tools is based on the parameterization of a single type of architecture, while another type employs a supporting library of lower-level primitives with which several types of architectures can be realized. With the advent of open systems a new generation of CAD tools was developed to allow multi-level iterations in the design process.

3.2.3 Component-level design

Once their specifications have been defined at the higher, functional building block level the design of circuit components can be carried-out in parallel by different designers according to their areas of expertise. This consists of defining a circuit component architecture together with the dimensions of the elements which are used for its instantiation. The verification of this level of design is usually carried out by traditional Spice-type electrical simulators.

3.3 Bottom-up flow for layout

The layout for integrated circuit implementation typically begins upon completion of the electrical design. It is based on a bottom-up flow of information whereby the basic elements are first laid-out; then, networks of those elements are assembled to layout a complete circuit component; then, several of those circuit components are interconnected to obtain a complete functional block; finally, various function blocks are put together according to the requirements of the higher, system level block architecture.

In the bottom-up layout flow, there are different types of difficulties addressing different design problems. For example, when a circuit component is laid-out the key design factors to be concerned with are the matching requirements of its basic elements and how these can be affected by the geometry format. When several of those circuit component layouts are assembled in order to form a complete functional block, the key layout factors are the block area, the routing density, the power dissipation distribution, the distribution of power lines and the parasitic capacitances at critical nodes. In addition, the layout should be devised to guarantee adaptation of the block to evolution of the topology, adaptation of the layout to evolution of the technology, the best definition of the inputs and outputs and, of course, a good component distribution and routing. Finally, the assembly of several functional building blocks to produce the layout of the complete macro block is constrained by the achievable block area, the routing density, the overall noise and cross-talk inside the block, the decoupling of analog and digital parts, the definition of inputs and outputs, the geometry format of the system and, of course, the adaptability for technology evolution and specification requirements.

3.4 CAD limitations

In the 70's, designs of integrated circuits were carried out in basically a single hierarchical layer taking into account just the physical layout of the circuit. This approach might have been reasonable for circuits with very limited complexity, merely of the order of tens of transistors. As soon as the complexity of the circuit increased it was clear that this methodology based on a flat layout approach was no longer adequate. Then, came the emergence of cell-based design methodologies and tools which allow the automatic or semi-automatic design of blocks based on libraries of pre-existing cells. Such tools have the flexibility of allowing the design of different types of architectures with varying levels of functional complexity, but they are restricted to the use of the cells available in a given library. Hence, in the present scenario of a fast evolving technology where there is little or virtually no time to consolidate libraries and where the requirements of the applications are also changing very rapidly, it is clear that such a design methodology is no longer appropriate.

Cell-based methodologies were later replaced by methodologies based on the automatic or semi-automatic compilation of functional building blocks. The CAD tools supporting such compilation methodologies are typically based on fixed architectures which merely allow the parameterization of the constituting elements according to the given requirements. However, because of the nature of high functionality block design and the complexity and variety of specifications that can be found in analog-digital interface blocks, the use of topology constrained solutions is also not satisfactory since it significantly reduces the number of solutions available for optimum design.

It is clear that in order to be able to cope with the requirements of evolution and the complexity of high functionality analog-digital blocks, such as those commonly found in interfacing applications, the design environment must be as open and as dynamic as possible and it should allow constant evolution and adaptability of both topologies and specifications. This represents an opportunity for the emergence of application-driven block design methodologies embedding intellectual property know-how. Indeed, the multidisciplinary nature and diversity associated with the design of a complete analog-digital interface block, where multiple functions are instantiated with multiple components, often leads to the uniqueness of a system solution attached to the know-how that produced it. Such application-driven block design methodologies are expected to meet the increased productivity of design that is vitally needed in the world semiconductor markets.

4. RETARGETABLE BLOCK DESIGN

4.1 Retargetable block model

The fundamental concept for the design of retargetable blocks is illustrated in Figure 6. After carrying-out a traditional top-down design, corresponding to the left hand side of the illustration, the results will be in the form of separate schematics and layout cells for the circuit components and functional building blocks. Each one of those schematics represents the connectivity and values of basic elements whereas the layouts represent their physical implementations and shapes for actual integrated circuit fabrication. In addition to this basic circuit information, there may also be experimental results of prototype chip characterization to validate the design solutions and possibly establishing their practical limits of effectiveness. We may now consider that all such information at schematic, layout and silicon levels can be embedded into a single block model. Such block model is clearly a very complex object that combines information from the electrical design, physical layout implementation and even from experimental characterization. Around each block model we can define a range of specifications which allow the fast retargetability of the model. Therefore, whenever a new set of specifications is defined within such range, the block model provides an important central platform from where the new block design can be quickly accomplished.

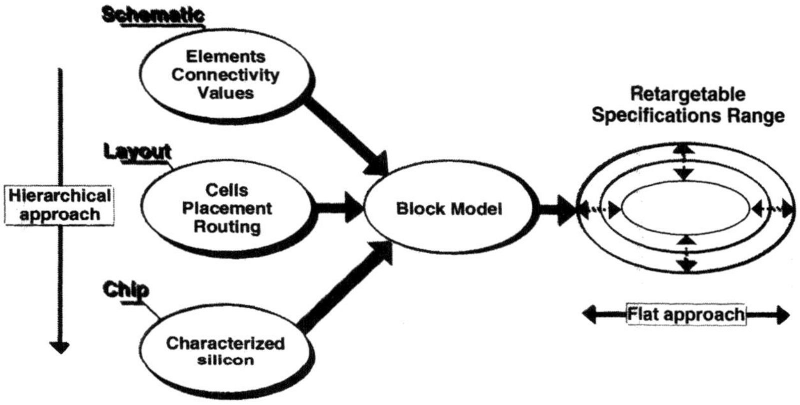

Figure 6. The concept of block model embeds information from electric schematics, physical layouts and even experimental characterization of a design.

In a CAD environment for retargetable block design, traditional libraries are replaced by "knowledge" libraries containing objects that always need some retuning design in order to meet new specifications. Rather than fixed designs, such objects are basically formed by design platforms which contain guidelines for determining the final and correct values of the elements. These can be expressions, tables or even simple computation routines.

4.1.1 Design flow for retargetability

The design flow to carry-out the retargeting of a block is illustrated in Figure 7. First, it considers the validation of a new set of specifications against the range of specifications for which the block has been previously designed. The design is carried-out only if the specifications are validated, meaning that the new set of specifications falls within the range of the retargetability of the block. In this case, a set of priorities are initially defined in order to define the sequence of specifications that should be accomplished first by the constituting functional building blocks according to their inherent adaptability and suitability for integrated circuit realization. Once such sequence is defined, the design process consists of tuning the existing designs to the new specifications. Then, after tuning the functional building blocks, there is a process of tuning the circuit component cells within each functional block and which consists also of changing only the parts of the circuit components that need to be changed. Suppose, for example, the retargeting of a block which is needed to increase the driving capability of an operational amplifier. It is clear that it would not be necessary to re-design the whole operational amplifier again, but simply to carry-out a much smaller intervention in the output stage, may be by adjusting its biasing level or re-sizing the dimensions of its constituting transistors.

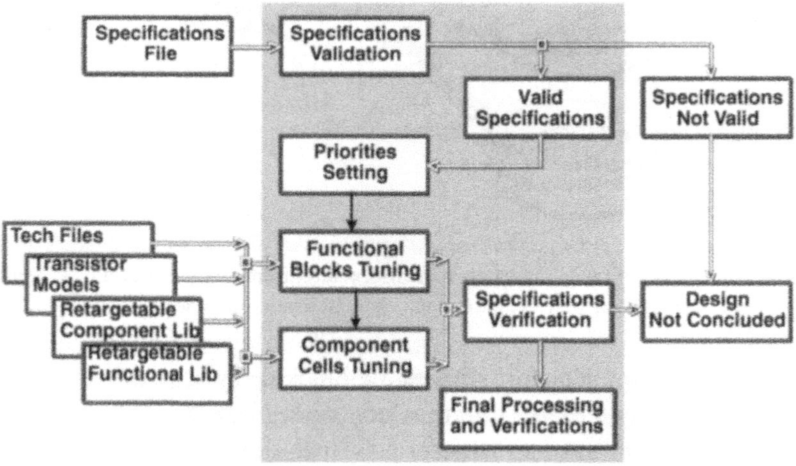

Figure 7. Typical flow in a retargetable block design environment.

After the tuning of all the functional building blocks and circuit components has been completed, it is necessary to verify the correctness of the new performance characteristics and then proceed to the final processing and verifications. If the target specifications cannot be met then the block design can not be concluded from the original block model, even if its initial specifications could have been validated. In principle, it is possible to define the range of retargetability so that the occurrence of unfinished designs is minimal. Nevertheless, by carrying out a tentative retargeting of a block based on assumptions of the validity of the specifications and by progressing through the different levels of the design, it is always possible to determine whether the design, indeed, cannot be concluded.

4.1.2 Key features of designing for retargetability

The required features for design quality and productivity in the above environment are multiple. First, it is necessary to ensure that the trading-off of specifications of the multiple functional blocks inside the block converges and that in such a process the system partitioning is optimized for implementation. Second, it is necessary to ensure that the block model is easily adapted to a technology evolution by using circuit component design methodologies which reduce the technology dependency. Third, it is required to ensure that the retargeted design achieves an effective use of silicon area because, in volume production, the area of the chip is paramount for determining the final production cost. This can be assured by maximizing the layout regularity, optimizing the component aspect ratios or even by establishing careful floor planning for the block. Finally, the routing density and routing area of the block should also be minimized as this represents an overhead area which is not directly needed for the functionality that needs to be delivered and, therefore, the higher the overall routing area, the higher the production cost of the circuit. This can be assured by a detailed analysis of the route-path characteristics and the use of simplifying routing techniques.

4.1.3 Design productivity

The productivity gains that can be achieved by using a methodology based on retargetable block designs is illustrated in Table 1. The original design of a given block is carried-out for specifications A, in 1.2 μm CMOS technology. This is carried out following a traditional systematic hierarchical approach starting with a system level architectural analysis and which is the process through which we define the complete architecture of the block. Since the target technology is being used for the first time, it is also necessary to develop a technology file as well as the library of the basic elements that are needed for instantiating the functional building blocks of the block architecture. This is usually a lengthy process that can take several months to complete depending on the block complexity.

Table 4. Illustration of the productivity gains achieved through retargetable block design methodologies.

Block Design	Analysis	Tech File	Schematics	Design Time
Specs A, 1.2 μm	New	New	New	8 months
Specs B, 1.2 μm	---	---	Modified	3 weeks
Specs A, 0.7 μm	---	New	Modified	4 weeks
Specs C, 0.5 μm	---	New	Modified	2 weeks

A new application requires designing the same block for specifications B in the same 1.2 μm CMOS technology. Assuming that such specifications are validated for the original block model, it is obvious that it is not necessary to carry-out the initial system level block analysis. Similarly, because the implementation technology is the same it is not necessary to develop a new technology file. However, since the new block specifications are different from the original ones it may be necessary to re-tune the functional building blocks and the constituting circuit components. This, of course, is a much faster process than having to re-design the whole block and can be completed in just a fraction of the time of the original design.

As it is also illustrated in Table 1, similar analysis of productivity gains can be made for different implementations of the same block, either in new technologies or for new specifications. In either case, the existence of a central design platform provided by a block model provides much faster design cycles than those achieved through traditional design flows. This will be illustrated next considering a number of practical design implementations of industrial products that have been developed based on such a methodology.

5. EXAMPLES FROM INDUSTRY PRACTICE

5.1 Quadrature D/A RF interface

The first example to be considered is the mixed-signal block discussed earlier in the chapter and which is required to provide the interface between two digital signals in quadrature, one in-phase and the other 90° out-of-phase, and the analog signals that interface an RF transmitter for wireless applications. For convenience of analysis, the corresponding block architecture is repeated in Figure 8. Each channel comprises a digital processing function that implements a finite impulse response filter together with a sinc cube filtering function and digital interpolation. This is followed by a steering-current digital-to-analog converter and an active-RC continuos-time filter providing both a current-voltage conversion and second order anti-imaging filtering function. Both quadrature channels are required to have very good matching, in terms of gain and phase, and the offset of each channel is also required to be very small. For this reason, the block includes an additional unit for gain and offset calibration and which, together with the overall block functionality, is controlled by a digital control block. Fully differential circuit techniques are obviously necessary to be employed throughout the block due to the stringent requirements for noise and power supply rejection.

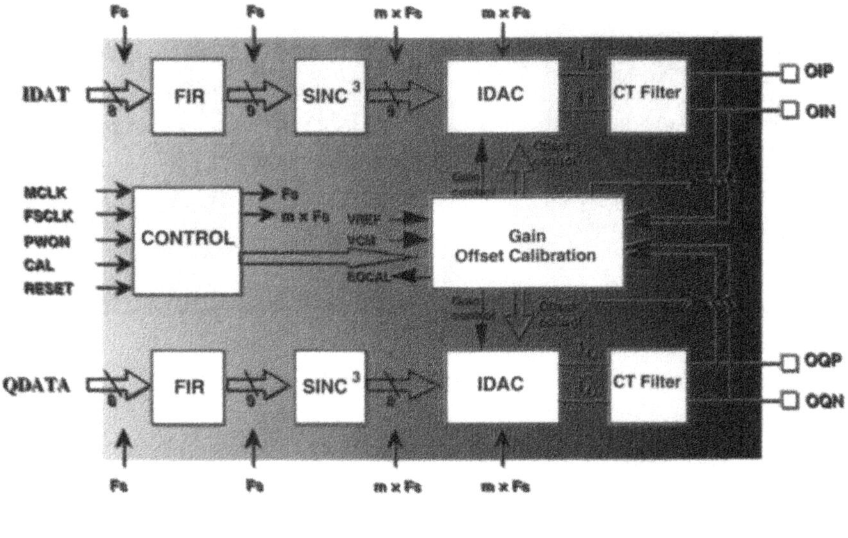

Figure 8. Block diagram of a mixed-signal block providing full interfacing functionality between a quadrature digital port and a quadrature analog port.

The above block has been retargeted for different applications, all for wireless communications, and which have different requirements for the digital input data, baseband frequency band, sampling frequency and even out-of-band image rejection. Two examples of the resulting layouts of such block with retargeted characteristics are illustrated in Figure 9. First, in Figure 9 (a), we can see the I&Q D/A interface for the Japanese PHS system, with 25 kHz baseband, and which required an 8-fold interpolation of the input digital signal, an 8-bits D/A converter and a 2nd order active-RC filter with nominal 400 kHz cut-off frequency. In Figure 9 (b) we can see the layout of the same I&Q D/A interface architecture retargeted for a GSM application, with 100 kHz baseband, and which employs a 10-fold digital interpolation, a 10-bits D/A converter and a 2nd order active-RC filter with nominal 800 kHz cut-off frequency. A careful inspection of the layouts of both blocks shows the differences in each one of the constituting block areas. For example, the output active RC filters which correspond to the large areas on the right-hand side of the layout; we can also see different areas for the steering current D/A conversion in the middle of the blocks and the different areas for the digital processing unit just before the D/A conversion functions. In the first case, we clearly see that signal processing function. In the second case, this is not included in core as the customer has decided to implement that function together with other digital signal processing functions.

(a)
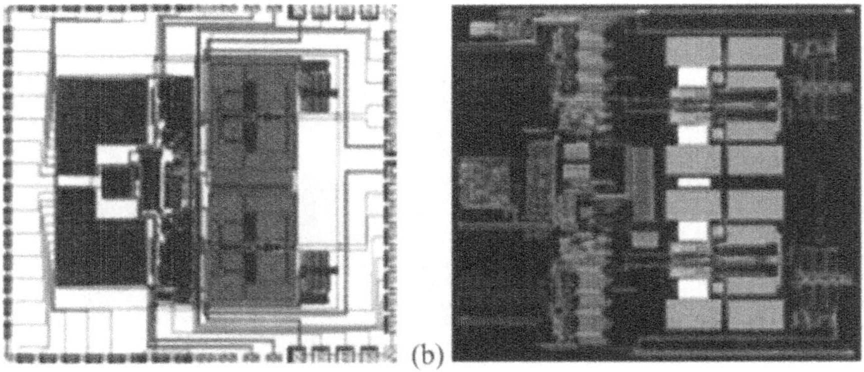
(b)

Figure 9. Retargetable block layouts based on the architecture of Figure 8.

5.2 Delta-sigma A/D interface

Another example of an analog-digital interface block, based on oversampling delta-sigma technology, is illustrated in Figure 10. It is formed by a programmable gain amplifier followed by a 2nd order single-bit delta-sigma modulator, which provides a high-speed bit stream representation of the analog input signal, and finally the digital lowpass filtering and decimation functions. These digital processing functions employ on binary descriptions of the filter coefficients in order to achieve highly compact physical implementations in integrated circuit form.

The above block has been retargeted for different characteristics to meet different application requirements. One such application is for voiceband digital processing with 4 kHz baseband while another application is for baseband radio processing, with 25 kHz baseband and complex I&Q processing. In Figure 11 we can see the layouts resulting from the retargeted design of the architecture of Figure 10. The layout in Figure 11 (a) corresponds to the single-channel voiceband application where the layout in Figure 11 (b) corresponds to the quadrature I&Q A/D interface which is complementary to the D/A interface described before.

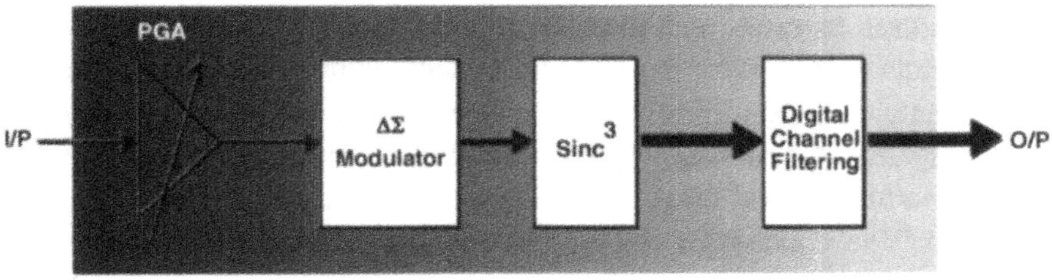

Figure 10. An analog-digital interface block based on oversampling delta-sigma technology.

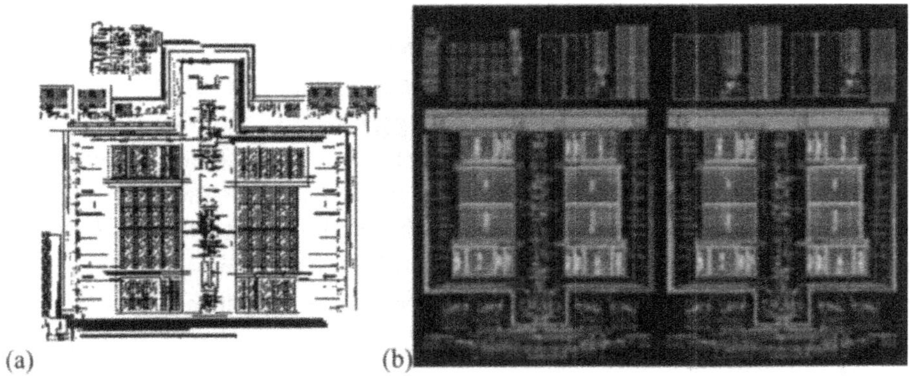

Figure 11. Retargetable block layouts based on the architecture of Figure 10.

The above examples have illustrated the use of application-driven retargetable analog-digital blocks to achieve increased design productivity. In both cases, the blocks were initially designed for a very specific set of specifications following systematic hierarchical methodologies for electrical and layout designs. Then, once the block models were captured, as discussed earlier, the retargetability for different applications has been carried-out very quickly. The resulting gains in productivity were obvious. If the same systematic top-down methodology had been used for independent re-designs of each one of those different interface blocks, the time spent would have been too long and market opportunities would have been lost. By contrast, the use of design retargetability methodology was instrumental to achieve a very quick time-to-market and hence providing cost effective solutions without compromising the quality and efficiency the design.

6. CONCLUSIONS

The material presented and discussed in this chapter makes it clear that mixed-signal integrated circuits will play an increasingly important roll in semiconductor markets and that the rate of technology and application developments will increase the pressure to cope with far greater productivity requirements for designing. Firstly, because of their intrinsic multi-disciplinary nature and secondly because of the lack of adequate tools that cross all the hierarchical levels of design, from architectural design to layout generation. The concept of retargetable, application-driven analog-digital blocks will provide the solution to those requirements in terms of design productivity and cost effectiveness. In order to achieve this, new methodologies and supporting tools encapsulating hierarchical multi-level design information must be developed alongside comprehensive behavioral models embedding both functional and electrical performance information. The examples that have been discussed at the end of this chapter, corresponding to practical industrial applications of the concept of retargetability, clearly demonstrate that very fast designs can be achieved without compromising the performance in terms of electrical operation, silicon area and power dissipation of those blocks. It is clear that the development of such methodology is still in its infancy and therefore significant efforts have to be deployed world-wide so that the tools that are needed to provide the required design support become widely available.

7. ACKNOWLEDGMENTS

The valuable contributions of Prof. Carlos Leme and Prof. João Vital for the development of the methodological aspects discussed herein are gratefully acknowledged. The integrated circuit design examples presented in this chapter are courtesy of ChipIdea Microelectronics Limited.

8. REFERENCES

F. Op't Eynde, et al., *Analog Interfaces for Digital Signal Processing Systems*, Boston: MA: Kluwer Academic, 1993.

G. Beenker, et al., "Analog CAD for consumer IC's", in *Analog Circuit Design*, Huijsing, van der Plassche, and Sansen Eds., Boston, MA: Kluwer Academic, 1993, ch. 15.

M. Degrauwe, et al., "IDAC: An interactive design tool for analog CMOS circuits", *IEEE J. Solid-State Circuits*, vol. SC-22, pp. 1106-1116, Dec. 1987.

P. Allen, et al., "A silicon compiler for successive approximation A/D and D/A converters", in *Proc. Custom Integrated Circuits Conf. (CICC)*, 1986, pp. 552-555.

G. Jusuf, et al., "A performance driven analog-to-digital converter module generator", in *Proc. Int. Symp. Circuits Syst. (ISCAS)*, 1992, pp. 2160-2163.

N. Horta, et al., "Framework for architecture synthesis of data conversion systems employing binary-weighted capacitor arrays", in *Proc. Int. Symp. Circuits Syst. (ISCAS)*, 1991, pp. 1789-1792.

N. Horta, et al., "Automatic multi-level macromodel generation for data conversion systems employing binary-weighted capacitor arrays", in *Proc. Int. Symp. Circuits Syst. (ISCAS)*, 1992, pp. 2561-2564.

G. Ruan, et al., "Modeling and simulation of ΔΣ A/D converters using a mixed-mode simulator", in *Proc. Custom Integrated Circuits Conf. (CICC)*, 1992, pp. 12.6.1-12.6.4.

F. Medeiro, et al., "A vertically-integrated tool for automated design of ΔΣ modulators", in *Proc. Euro. Solid-State-Circuits Conf. (ESSCIRC)*, 1994, pp. 164-167.

J. Vital, et al., "Synthesis of high-speed A/D converter architectures with flexible functional simulation capabilities", in*Proc. Int. Symp. Circuits Syst. (ISCAS)*, 1992, pp. 2156-2159.

G. Casinovi, et al., "A macromodeling algorithm for analog circuits", *IEEE Trans. Computer-Aided Design*, vol. 10, pp. 150-160, Feb. 1991.

L. Williams, et al., "MIDAS - A functional simulator for mixed-signal digital and analog sampled-data systems", in *Proc. Int. Symp. Circuits Syst. (ISCAS)*, 1992, pp. 2148-2151.

Chapter 6

Robust Low Voltage Low Power Analog Mos VLSI Design

Tuna B. Tarim[1] , Chi-Hum Lin[2], ,Mohammed Ismail[3]
[1] *Mixed Signal Wireless Division at Texas Instruments, Inc., Dallas, Texas, USA email: tuna@ti.com*

[2] *Analog VLSI Lab, The Ohio State University, Columbus, OH, USA, e-mail: linc@ee.eng.ohio-state.edu*

[3] *Analog VLSI Lab, The Ohio State University, Columbus, OH, USA and Radio Electronics Lab, Royal Institute of Technology, Kista-Stockholm, Swede,; e-mail: ismail@ee.eng.ohio-state.edu*

Keywords Statistical design, yield enhancement, design of experiment, response surfaces, low voltage, low power, rail-to-rail.

Abstract: Integrating analog, mixed signal and RF circuits with digital in a System-on-Chip (SoC) design solution is a major trend nowadays and finds many applications in areas like wireless and wireline communications and multimedia applications.

This chapter presents statistical design techniques leading to optimization and yield enhancement of integrated CMOS analog and mixed signal solutions. In a SoC design, minimizing yield loss that often results from incorporating analog or RF parts in a large SoC digital design is becoming increasingly important to maintain a cost effective total solution. This is particularly true in today's deep sub-micron technologies where random process variations, supply noise and ground bounce become increasingly critical .Robust design techniques at both the schematic and physical layout levels will be discussed and demonstrated with design examples of low voltage CMOS analog integrated circuits .

1. INTRODUCTION

This Chapter focuses on robust design of low power CMOS analog VLSI circuits. The need for such circuits is tremendously growing [1]. In order to produce manufacturable analog integrated circuits with a high functional yield and a high degree of reliability, the design of such circuits must be robust with respect to random process and device parameter variations. This is particularly true when analog and mixed analog/digital VLSI circuits operate from a low supply voltage [2-4] (3V and below) due to the fact that random variations do not scale down with feature size or supply voltage. Such variations, therefore, could ultimately be a limiting factor on how low could be [5]. The chapter will discuss basic design techniques to circumvent this problem at both the circuit schematic level (e.g., aspect rations of CMOS devices) and the physical layout level (e.g., device areas and separation distances on chip0. Section 1 will introduce low voltage low power square-law CMOS composite transistors. Section 2 describes statistical VLSI tools and techniques used in the robust design of analog ICs by incorporating random variations into the design flow. Section 3 discusses the robust design of the composite cells introduced in Section 1. Section 4 presents a robust design technique at the schematic level for low voltage CMOS operational amplifiers and Section 5 discusses the robust design of such operational amplifiers at the physical layout level. Design examples are included throughout the chapter to reinforce understanding of the basic concepts.

2. LOW VOLTAGE CMOS SAQUARE-LAW COMPOSITE CELLS

A single CMOS transistor (see Fig.1) exhibits the square law dependency on the gate-source voltage required in numerous signal processing applications [6-10}. However, the low input impedance at the source of the transistor limits the applicability of the single transistor solution and calls for more subtle designs in many cases. A simple solution would be to use two-transistor CMOS composite cell proposed by Seevinck an Wassenaar [8] given in Fig. 1. It exhibits the same square-law behavior of a single transistor with high input impedances at its equivalent gate (V_g) and equivalent source (V_S) terminals. However, it is not suitable for low voltage designs due to its high equivalent threshold voltage, $V_{Teq} = | V_{Tp} | + V_{Tn}$. In this section, new low voltage composite cells with two high impedance terminals which control the current flow through the transistors are introduced. The cells are show in Fig. 1.

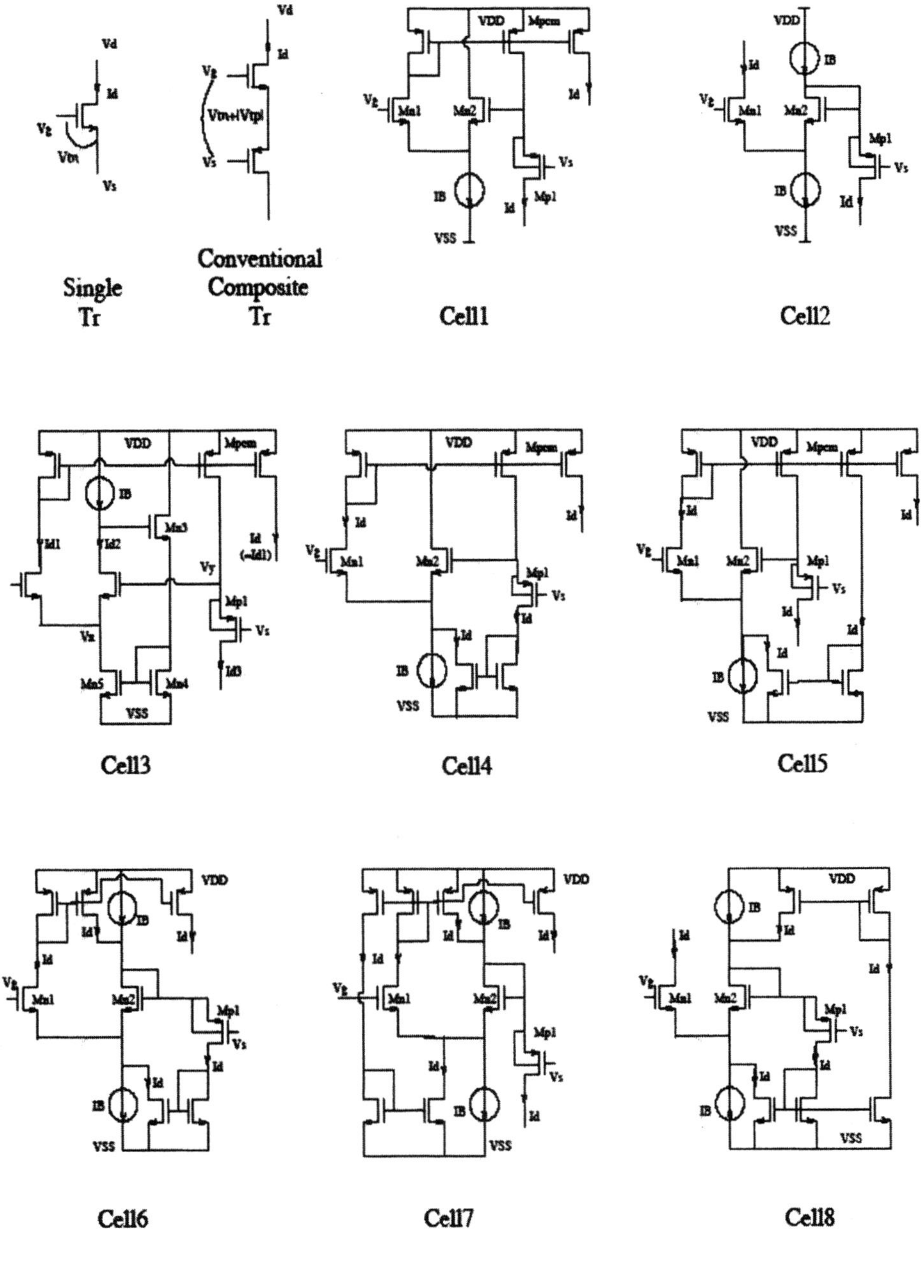

Figure 1. Single, conventional and low voltage composite cells

*Cell*1 [1] and *Cell*2 [2] in Fig. 1 have the drain current Id expressed as

$$Id = \frac{K_{eq}}{2}\left(V_{gs} - V_{Teq}\right) \qquad (1)$$

Where Keq and V_{Teq} are the equivalent tranconductance parameter and the threshold voltage, respectively, given by

$$Keq = \left(\frac{1}{\sqrt{K_{n1}}} + \frac{1}{\sqrt{K_{p1}}}\right)^{-2} \qquad (2)$$

$$VTeq = \left|VTp\right| - \sqrt{\frac{2(I_B - Id)}{Kn2}} \qquad (3)$$

In the case where $I_B > I_d$, equation (3) can be simplified as

$$VTeq = \left|VTp\right| - \sqrt{\frac{2I_B}{K_{n2}}} \qquad (4)$$

$K_{eq,}$ given by equation (2) is the as that of the conventional two-transistor composite cell. However, V_{Teq} given by equation (4) is of the conventional composite transistor [8] and hence is suited for low voltage applications. Modifications [9] of *Cell*1 and *Cell*2 to reduce power consumption are discussed next.

As described previously, both circuits should satisfy the assumption the $I_B > I_d$ to reduce V_{Teq} variation in equation (3) , otherwise it could degrade the accuracy of the signal. Therefore, I_d has to be set at a much larger value than the maximum I_d implying that I_d can not be increased more than $I_{B.}$ This causes high power consumption at quiescent conditions due to the large $I_{B.}$ This problem can be overcame by using the following methods:

1. A constant voltage shifter can be used (*Cell*s based on *Cell*1): the feedback loop formed by transistors $M_{n2,}$ $M_{n3,}$ $M_{n4,}$ and $M_{n5,}$ and the bias current I_B. This makes the voltage drop V_{gs2} (or $V_{gs2} - V_{Tn}$) of M_{n2} constant.
2. An adaptive bias technique can be used (*Cell*4 and *Cell*5 based on *Cell*1 and *Cell*6, *Cell*7 and *Cell8 based on Cell*2):

in order to cancel the I_d effect in equation (3), the current $I_B + I_d$ can be used instead of the constant current I_B. For this purpose, a current mirror circuit is used to copy I_d and add it to I_B. The summed current, $I_B + I_d$, is replace with I_B in the second term of the right hand side in equation (3). Thereby, equation (3) becomes equation (4), which is constant regardless of the input voltages applied.

The proposed circuits have been simulated using the MOSIS 2 μ m well process with V_{Tn} =0,82V,and a supply voltage of 3V . The drain currents of each cell (*Cell*4, *Cell*6 and *Cell*8) with I_B=120μ A, a single NMOS transistor with K_n and V_{Tn} equivalent to K_{eq} and V_{Tn} equivalent to K_{eq} and V_{Tn1} (equation (4)), respectively and numerical evaluations of equation (1) with K_{eq} and V_{Teq} (equation (4)) are shown in Fig. 2 (a) as a function of V_g, with V_S =0.5V, where each cell exhibits a square-law drain current characteristic on V_{gs}. A similar simulation is Performed with I_B =12 μ A , in Fig. 2(b). It should be noted in this figure that the drain currents of *Cell*3-8 can increase beyond I_B (=12 μ A). This is not possible with the first two cells, *Cell*1 and *Cell*2.

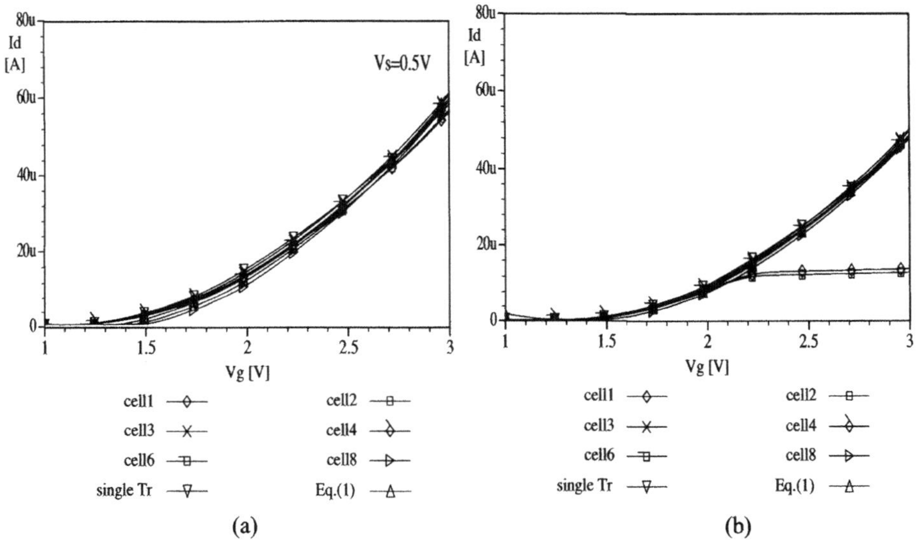

*Figure 2.*DC characteristics of *Cell*1-8 for (a) I_B=120μ A and (b) I_B=12μ A

3. STATISTICAL VLSI DESIGN TOOLS AND TECHNIQUES

As feature sizes in MOS processes move into the submicron range and power supply voltages are reduced, the effect of both device mismatch and inter-die process variations on the performance and reliability of analog circuits is magnified. In order to fully utilize the capabilities of a given process, a circuit designer needs to have both, a complete knowledge of the statistical distributions of transistors parameters produced by the process and a way to determine the effects of variations in these parameters on circuit performance. To produce cost-effective, manufacturable analog and mixed-signal chips, circuit designers must be used. This is even more critical for submicron low voltage designs since random variations do not scale down with feature size or supply voltage. Moreover, with current trends of higher levels of integration leading to complete mixed-signal systems on a chip, yield loss due to the analog part must be minimized such that it has little effect on the yield of the mixed-signal chip [11].

Previous studies in the area of statistical modeling and simulations were separately successful in determining both functionality of parameter mismatch variance and a methodology to simulate circuits, but merging the results of these two fields into a unified method to model and simulate performance variances in circuits containing MOS devices is a recent study [12].

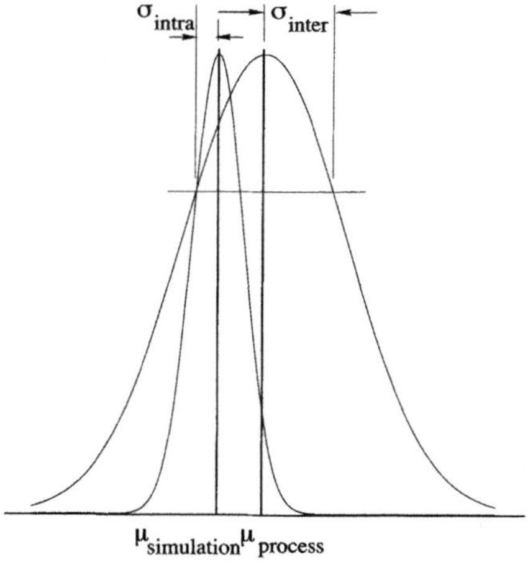

Figure 3. Relationship between inter-and intra-die parameter standard deviations

The statistical device model must describe the correlation between parameters, as well as the parameter means and variances. Since analytical models are costly to build and difficult to apply, statistical techniques have to be employed to build an empirical model. Some of these statistical techniques will be reviewed in this section.

3.1 Statistical Parameter Modeling

Random variations in integrated circuit processes cause random variations in transistor parameters. Causes of circuit output variance can be divided into two groups [12]: inter-die device variability and intra-die device variability. Inter-die variability is characterized by die-to-die, wafer-to-wafer, or lot-to-lot process variability. Since inter-die variability equally affects all transistors in a given circuit, it can be represented by a deviation in the parameter mean of every transistor in the circuit. The relation between inter-die parameter standard deviations are show in Fig. 3.

μ_{process} is the process-level parameter mean, and $\mu_{\text{simulations}}$ the randomly determined value of parameter mean used for each transistor in one circuit simulation. Inter-die parameter standard deviation is usually much larger than intra-die parameter standard deviation; however, in many analog circuits, it is the intra-die parameter variances or device mismatch which causes the greatest deviations in circuit performance. Intra-die device mismatch causes similarly designed transistor under equivalent biasing conditions to behave differently. This type of mismatch arises from wafer-level process variability and therefore is much more difficult to model. The degree of mismatch between two devices is, in general, dependent on both the size and relative location of the devices.

In analog integrated circuits, device mismatch contributes greatly to variances in circuit performance. The effect of inter-die device variability may be counteracted in many analog circuits either by automatic tuning techniques or by altering available bias conditions. Therefore, a statistical model which comprehends device mismatch is necessary for statistical simulations of analog circuits.

3.2 Parameter Variance Models for MOS Device Mismatch

In order to include random mismatch between circuit devices, the statistical model have a different set model parameters for each transistor in a circuit. The variance of the parameter mismatch between two transistors increases witch decreasing gate and increasing separation distance between device, which means, small, widely spaced devices have a statistically larger mismatch than large, closely spaced devices. Pelgrom and others [13-15] showed that the variance of the mismatch between two transistors can be represented by

$$\sigma^2(P_1 - P_2) = \frac{a_p}{2W_1L_1} + \frac{a_p}{2W_2L_2} + s_p^{\ 2} D_{12}^{\ 2} \tag{5}$$

Where D_{12} is the distance between transistors 1 and 2, W_1L_1 *and* W_2L_2 are the gate areas of transistors1 and 2, respectively, and a_p and S_p are process dependent fitting constants. This model is derives for a general parameter, and therefore should be valid any set of MOS transistor parameters. The model considers two of the greatest effects on device variability of analog circuits: device size and circuit layout. The three terms on the right-hand side of equation (5) are independent and are assumed to be normally distributed, which means , they can be accounted for separately when calculating transistor parameter sets in a Monte Carlo Scheme.

The MOS transistor is modeled by multiple parameters and these parameters are not independent. The correlation between variances of different parameters have to be preserved for a statistical model to be useful. One way of modeling the correlations between parameters in to use Principal Component Analysis (PCA) [12]. PCA is used to generate normalized parameters from independent unit random variables, using a series of linear equations. Thus, correlation among parameters is preserved by their PCA coefficients and the independent variables.

The statistical MOS (SMOS) model was development at The Ohio State University, Solid-State Microelectronics Laboratory [12]. This model has been incorporated into APLAC [16] and applied to the BSIM MOS model parameters because of the ability of the BSIM model to account for effects of both device geometry and biasing. The model calculation procedure for the SMOS model is given in Fig. 4.

The experimental work necessary to determine the model fitting constants is shown in the "process characterization" block. PCA preserves the parameter correlation information and the distance dependence of the mismatch is preserved with the of σ-space analysis. The XY layout coordinates of each transistor in the circuit is given in the "circuit description" block. The Monte Carlo analysis is performed in the inner loop, thus, the drawback of the requirement of excessive CPU time no longer exists for the SMOS model. The statistical circuit analysis allows for optimization of device sizes and separation distances to improve a certain performance criteria and for parametric yield estimation. With the aid of the SMOS model it is presently possible to simulate random device mismatch effects on the circuit performance.

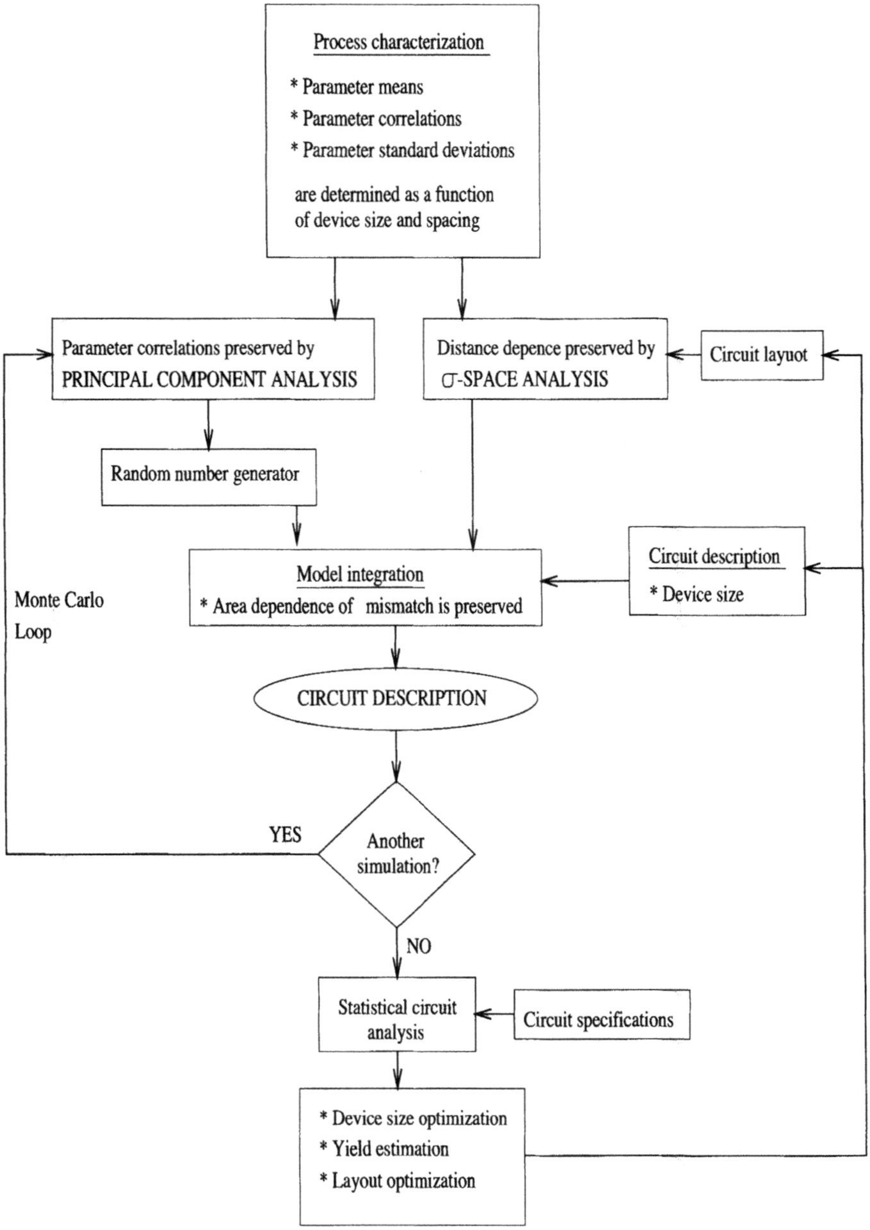

Figure 6.4. The model calculation procedure for the SMOS model [12]

3.3 Statistical Techniques

An accurate mathematical model must be constructed to quantitatively describe the circuit performance to be optimized. An analytical model is costly to build and difficult to apply. Therefore statistical techniques have to be employed to build empirical model. Response Surface Methodology (RSM) is one of the powerful tools to characterize the relationship between the output and independent input variables of a process by using a polynomial fitting model [17]. The input variables could be the areas of MOS transistors in the circuit. In order to reduce complexity, it is preferable to have as less input variables in the empirical model as possible. This can be done by determining the most effective input variables on the circuit performance to be optimized. The construction of the empirical model starts and continues with running a series of experiments. Design of Experiments (DOE) [18] is a widely used systematic method for experiment planning. Fig. 5 gives the steps of the DOE method.

The construction of the empirical model starts with running a series of experiments at different input variable levels. To apply DOE to a circuit modeling task, the designer should understanding the problem to be solved and recognize the input variables that affect the response under study. Based on the number of input variables, experimental coast and accuracy requirement of the model, the level and the method of experiment can be chosen. The level of an experiment is the number of values each input variable will be assigned to during the experiment. Usually, a two-or three-level experiment is used for most tasks.

The design of an experiment with multiple variables can be simplified by running two sets of experiments. The first run is a two-level screening test to determine the most contributing input variables. The Placket-Burman design was chosen for its simplicity [18]. The effect of each input variable V_i is indicated by the sun of square as where +1 and –1 represent the two, low and

$$SS(V_i) = \frac{N}{4}\left[avg(V_i = +1) - avg(V_i = 1)\right]^2 \quad (6)$$

high , levels.

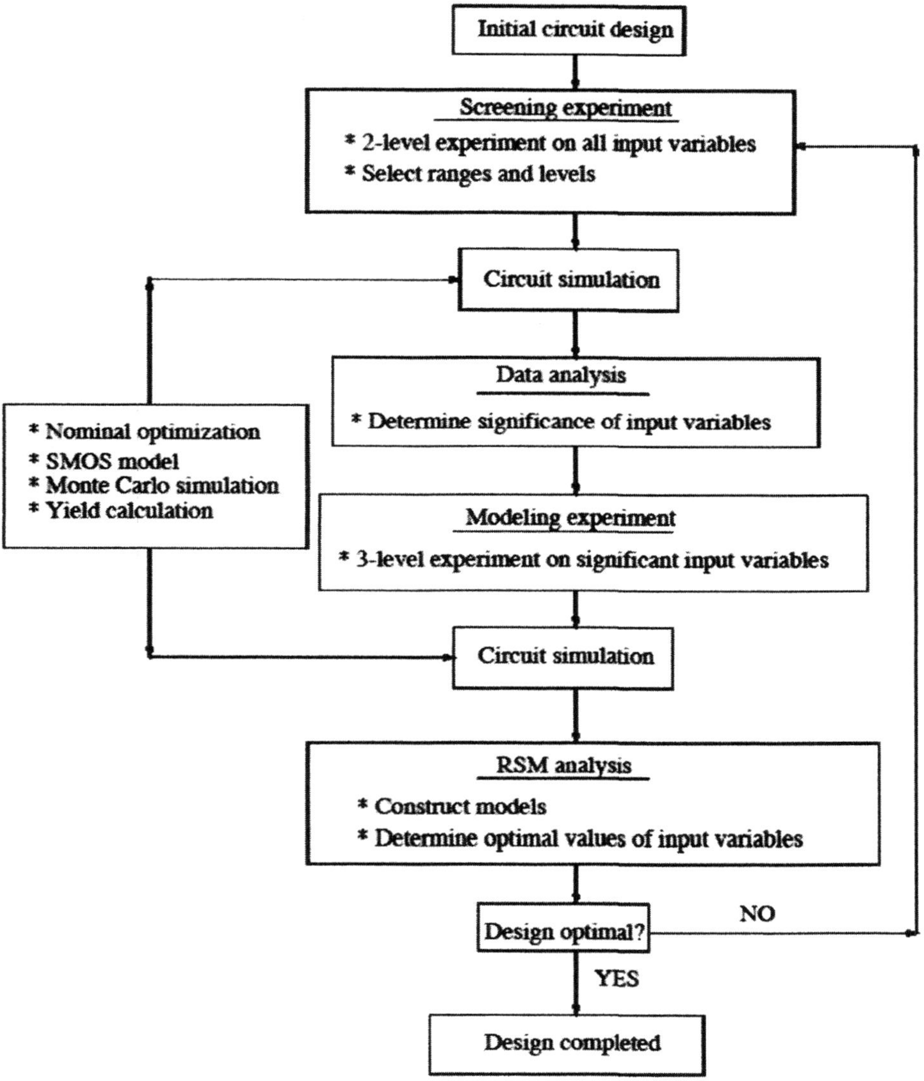

Figure 5. The design of experiments (DOE) algorithm

where –1 and +1 represent the minimum and maximum area values, respectively, and N is the number of runs. The significant variables whose SS values constitutes 95% of the total SS are considered in the second step of design which is a three-level model building experiment. A Box-Behnken design is suitable for this task, due its ability to construct a full quadratic model [18].

The three levels –1, 0 and +1 are the minimum, center and maximum values for areas that are carried over to the three-level Box-Behnken experiment, respectively. The results of the Box-Behnken design are analyzed and fitted to a polynomial model using the regression method. The regression method fits the data into a polynomial equation with the least squares, and the empirical model is constructed. The fitness of the model is indicated by the coefficient of determination R with;

$$R^2 = \frac{\sum\left(\hat{Y} - \bar{Y}\right)^2}{\sum\left(\hat{Y} - \bar{Y}\right)^2} \tag{7}$$

Where Y is the mean of all the observations an Y_i is the fitted value. A perfect fit should have $R^2 = 1$. Response Surface Methodology [13] is implemented into Minitab [19], and helps obtain the response surfaces used to find the optimal device area which achieves the desired circuit performance. The reliability of the regression model coefficients b_i is subject to the following hypothesis test:

$$H_o = b_i \tag{8}$$

$$H_o \neq b_i \tag{9}$$

where α is the level of significance. That t value

$$t = \frac{b_i}{s\{b_i\}} \tag{10}$$

indicates the credibility of a model coefficient, where $s\{b_i\}$ is the square root of the variance. The polynomial term with t value between –0.5 and 0.5 is considered unreliable and ignored.

4. STATISTICAL DESIGN OF THE CMOS SQUARE-LAW CMOS CELLS

This subsection contains examples for the statistical design and simulation of *Cell*1 and *Cell*3, given in Fig. 6.1. The statistical design and analysis methodology has been previously discussed in detail. The target for the example circuits will be to calculate and minimize the relative drain current mismatch.

To apply the SMOS model to the cells, layout information must be given. The X-Y coordinates of each transistor in the circuits is specified in the simulation program. The placement of the transistors are extracted directly fro m the actual layout of the cells.

Device mismatch is a function of the device area and the separation distance. The effects of the channel area, a, and the aspect ratio, b, are separated by including the following equations in the netlist, during the statistical simulation:

$$W = \sqrt{ab} \tag{11}$$

$$L = \sqrt{\frac{a}{b}} \tag{12}$$

4.1 Example 1: Statistical Simulation of *Cell*1

Our first example will be the statistical design and simulation of *Cell*1 [20]. We will try to minimize the current mismatch between the drain current of transistor M_{n1} and current I_{d1}. We will run the Placket-Burman screening experiment in order to find the SS values and try to understand which input variables are most effective variables, namely. a_{n1}, a_{n2}, a_{pcm} ,a_{p1} and a_{ncs} are selected for the cell, therefore, eight runs is required for the Placket-Burman experiment [14]. Table 1 shows the area level and assignments for each input variable.

Table2 shows the Placket-Burman design matrix. The last column in the table shows the experimental results for each run where I_{cm} is the value for the standard deviation of the current mismatch. Each run

Table1. [Area level and assignments for each transistor in *Cell*1]

Transistors	M_{n1}	M_{n2}	M_{pcm}	M_{p1}	Current source (I_B)
Area Symbol	a_{n1}	a_{n2}	a_{pcm}	a_{p1}	a_{ncs}
(-1) μ m^2	10	200	450	30	114
(+1) μ m^2	50	1000	2250	150	570

In the design matrix consists of a Monte Carlo loop of n=500 and takes about 10 seconds of CPU time on a HP 715/133 Workstation. The SS values for each variable are given in Table 3.

Table 2. [Plaket-Burman design matrix and results]

Run	a_{n1}	a_{n2}	a_{pcm}	a_{p1}	a_{ncs}	σ (I_{cm})
1	1	-1	-1	1	-1	0.0167
2	1	1	-1	-1	1	0.0222
3	1	1	1	-1	-1	0.0213
4	-1	1	1	1	-1	0.0278
5	1	-1	1	1	1	0.0144
6	-1	1	-1	1	1	0.0246
7	-1	-1	1	-1	1	0.0248
8	-1	-1	-1	-1	-1	0.0256

The results shows that transistors M_{n1}, M_{n2} and M_{p1} are the most influential transistors, contributing more than 95% of their respective total SS. Transistor M_{p1}, having the lowest contribution may not have an important effect on the results, however, it will be still included in the next step of the methodology, to be further investigated.

The second step of the statistical experiment is a three-level model building experiment, -1, 0, +1 representing the three levels of the design Variables.

Table 3. [SS values and contribution of each transistor]

Factors	SS	%
a_{n1}	9.9073 x 10^{-5}	69.61
a_{n2}	2.5975 x 10^{-5}	18.25
a_{pcm}	8.0244 x 10^{-8}	0.06
a_{p1}	1.3707 x 10^{-5}	9.63
a_{ncs}	3.4847 x 10^{-6}	2.45

The Box-Behnken design matrix for 15 runs is gives in Table 4, the last column giving the experimental results foe each run.

Table 4 is applied to the statistical program Minitab [13], and an empirical model for *Cell* 1 is obtained as

$$\sigma\,(I_{cm}) = 0.2487 - 0.0298a_{n1} + 0.00081a_{n2} + 0.02401a_{n1}^2 - 0.00009\,a_{n2}^2 + 0.00145\,a_{n1}a_{n2} + 0.00033\,a_{n2}a_{n1} \tag{13}$$

with 93.3% accuracy, where I_{cm} is the current mismatch is the performance under consideration a_{n2} , a_{n1} and a_{p1} are the areas of the most contributing transistors, and are also the variables of the empiral model. The limits for test variables are [10 μ m^2 , 50 μ m^2], [200 μ m^2 , 1000 μ m^2] and [30 μ m^2 ,150 μ m^2], respectively. Note that, only one term in the empirical model includes the term a_{p1}, which represents the area of transistor M_{p1} , thus, it is likely that this transistor may not have an important effect on matching. It is easy to see this when comparing the contribution of transistors, in Table 3, where transistors M_{p1} and M_{p2} have a higher contribution.

Fig. 6 shows the contour curves for a_{n1} vs. a_{n2} and a_{n1} vs. a_{p1}, respectively. The area of transistor M_{p1} and M_{n2} is kept constant, respectively, for both response surfaces.

There could be two ways to interpret and/or to make use of Figure 6:

1. If there is specific value that is preferred for each transistor, it is possible to find those values from the x and y axis, and find the intersection point. The value of the surface which crosses that intersection point gives the
2. standard deviation value of the current mismatch.

Table 4. [Box-Behnken design matrix and results]

Run	a_{n1}	a_{n2}	a_{p1}	σ (I_{cm})
1	-1	0	0	0.0248
2	1	0	0	0.0187
3	-1	0	0	0.0247
4	1	0	0	0.0233
5	0	0	0	0.0239
6	0	-1	-1	0.0199
7	0	-1	-1	0.0229
8	0	1	1	0.0192
9	0	1	1	0.0253
10	0	0	0	0.0233
11	-1	0	-1	0.0243
12	1	0	-1	0.0223
13	-1	0	1	0.0265
14	1	0	1	0.0241
15	0	0	0	0.0230

1. If there is a certain current mismatch that is preferred , e.g., according to the design specifications, the circuit cannot tolerant more than a certain value of current mismatch, it is possible to find the surface that corresponds to that value. Then, the areas that inter-sect on that solution. Obviously, there will be more than one solution on the same surface; this brings the preferred flexibility of selecting the suitable area values for different designs.

The x and y axis shows area values for the transistors. The designer will select the appropriate W and L value which gives that area. The lowest standard deviation value which can be achieved from the curves in Fig. 6 is 2.1% with the appropriate sizing of the transistors. The designer has the flexibility of deciding if those values are appropriate for the circuit, and if not, to be prepared for the variation on the performance.

It is also possible to use the standard deviation information to enchance the yield. Let us assume that the goal of optimization is to obtain the minimum device area while achieving $I_{cm} \langle 5\%$, with a functional yield of 95%, or equivalently, to achieve the standard deviation of the relative drain current mismatch of 2.5%, since 95% is approximately $\pm 2\sigma$. From Fig. 5(a), the minimum point on the response surface corresponding to 2.5% is found to be $a_{n1} = 10\,\mu\text{m}^2$, $a_{n2} = 300\,\mu\text{m}^2$, and $a_{p1} = 90\,\mu\text{m}^2$. From the definition of W and L given in Section 3 $(W/L)_{n1} = 5/2$, $(W/L)_{n2} = 122/2.5$, and $(W/L)_{p1} = 26/3.5$. Thus, when these aspect ratios are used for the three transistors, the standard deviation will not exceed 2.5%, and the functional yield will be 95%. The only way to prove this result is to fabricate the circuit in large numbers and calculate the standard deviations from the measurement data. However, the whole purpose of making statistical design is to able to estimate the yield and standard deviation without actually having to fabricate the circuits, in order to reduce the cost. Statistical analysis results will give insight to the designer, and a quantitative measure of how much the standard deviation will be for the circuit performance under consideration.

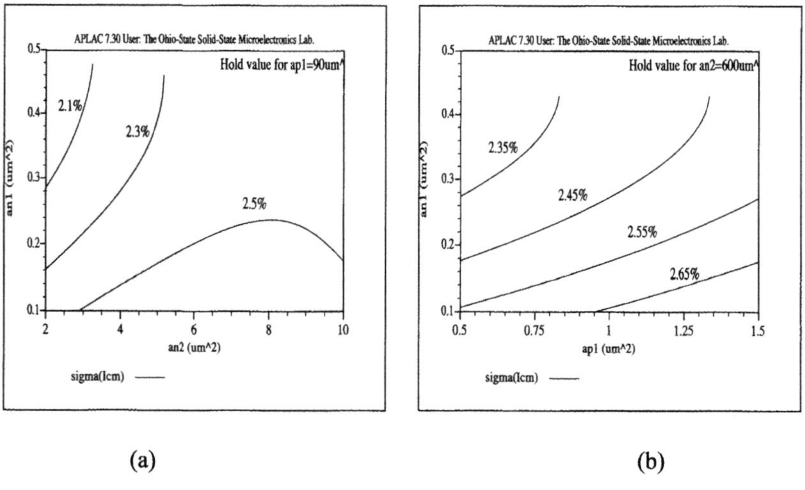

(a) (b)

Figure 6. Response surfaces for (a) a_{n1} vs. a_{n2} ($a_{p1} = 90\,\mu\text{m}^2$), (b) a_{p1} ($a_{n2} = 600\,\mu\text{m}^2$).

4.2 Example 2: Statistical Simulation of *Cell3*

Cell3 is the second composite cell to be examined [20]. As discussed in the previous example, a two-step experimental procedure can be used in the robust design of the cell. Seven input variables, namely a_{n1}, a_{n2} a_{pcm}, a_{n3}, a_{p1} , a_{n4n5}, and a_{pcs} are selected for the first of the experiments. Table 5 shows the area and level assignments for each variable in the circuit.

Table 5. [Area level and assignments for each transistor in *Cell1*]

Transistors	M_{n1}	M_{n2}	M_{pcm}	M_{n3}	M_{p1}	M_{n4n5}	M_{pcs}
Area Symbol	a_{n1}	a_{n2}	A_{pcm}	a_{n3}	a_{p1}	a_{n4n5}	a_{pcs}
(-1) μ m^2	10	200	450	1000	30	114	450
(+1) μ m^2	50	1000	2250	5000	150	570	2250

Since the number of input variables are seven, eight runs for the Placket-Burman design is sufficient. The Placket-Burman design matrix for eight runs and seven variables is given in Table

The last column of Table 6 shows the results of the eight runs of Placket-Burman simulations. Each run in the design matrix takes 14 seconds of CPU time on a HP715/133 workstation. Table 7 shows the SS values and the contribution of each transistor to the circuit performance.

Table 6. [Placket-Burman design matrix and results]

Run	a_{n1}	a_{n2}	a_{pcm}	a_{n3}	a_{p1}	a_{n4n5}	a_{pcs}	σ (I_{cm})
1	1	-1	-1	1	-1	1	1	0.0239
2	1	1	-1	-1	1	-1	1	0.0283
3	1	1	1	-1	-1	-1	-1	0.0248
4	-1	1	1	1	-1	-1	1	0.0293
5	1	-1	1	1	1	-1	-1	0.0204
6	-1	1	-1	1	1	1	-1	0.0327
7	-1	-1	1	-1	1	1	1	0.0291
8	-1	-1	-1	-1	-1	-1	-1	0.0295

Table 7 . [SS values and the contribution of each transistor]

Factors	SS	%
a_{n1}	6.7192×10^{-5}	62.27
a_{n2}	1.8486×10^{-5}	17.13
a_{pcn}	1.5141×10^{-5}	14.03
a_{n3}	3.5175×10^{-6}	3.26
a_{p1}	1.0883×10^{-6}	1.01
$a_{n4,n5}$	1.1173×10^{-6}	1.04
a_{pcs}	1.3660×10^{-6}	1.27

It is seen that four transistor contributions , namely, M_{n1}, M_{n2}, M_{pcm} and M_{n3} , will sum up to 95%. However, the contribution of the fourth variable is small compared to the other three.It is also experience from the previous circuit that a variable with small contribution is not going to be effective on the contour curves, hence, it will suffice to take the first three variables and include them in the step.

Table 8 shows the Box-Behnken design matrix and simulation results. Each run in the design matrix takes 14 seconds of CPU time on a HP 715/133 workstation.

The Box-Behnken design matrix was applied to the statistical saftware Minitab, to obtain the empirical model for the circuit. The resulting model is a given as

$$\sigma\,(I_{cm}) = 0.3451 - 0.04611a_{nl} + 0.00056a_{n2} - 0.00031a_{pcm} + 0.4281a_{nl}^{\,2} - 0.00008a_{n2}^{\,2} + 0.00120a_{nl}a_{n2} + 0.0003a_{n2}a_{pcm} \qquad (14)$$

with a 93.9% accuracy, I_{cm} , in equation (14) is the current mismatch . The 'T" value for the term $a_{n1}a_{pcm}$ is between the values –0.5 and 0.5, hence, this term is excluded from the empirical model. The coefficient of a_{pcm2} is "0", so this term is not in the model either. The limits for the variables a_{n1}, a_{n2} , a_{pcm} are [10 μ m^2, 50 μ m^2], [200 μ m^2 , 1000 μ m^2]

Table 8. [Box-Behnken design matrix and results]

Run	a_{n1}	a_{n2}	a_{pcm}	$\sigma\,(I_{cm})$
1	-1	-1	0	0.0286
2	1	-1	0	0.0195
3	-1	1	0	0.0313
4	1	1	0	0.0259
5	0	0	0	0.0258
6	0	-1	-1	0.0257
7	0	1	-1	0.0253
8	0	-1	1	0.0224
9	0	1	1	0.0265
10	0	0	0	0.0261
11	-1	0	-1	0.0302
12	1	0	-1	0.0260
13	-1	0	1	0.0303
14	1	0	1	0.0253
15	0	0	0	0.0257

And [450 μ m^2 , 2250 μ m^2], respectively; the empirical model is valid within these ranges.

Figure 6 gives the contour curves for different pairs of variables. It is possible to make use of the curves the same way as explained in the previous section. Never to mention, the factor a_{pcm} seems to be insignificant for this circuit, thus the main focus is on a_{n1} and a_{n2}. The standard deviation is in the range of 2.5%, some what higher than the previous circuit, yet still comparable.

The low voltage low power composite cell has a feedback loop and also the number of transistors are more than the low voltage composite cells, thus, the circuit is more complicated. However, the standard deviation does not differ too much. It is possible to conclude that the circuit not only avoids the rate-off between low voltage operation and low power dissipation, but statistically speaking, overcoming the trade-off does not cost anything to the circuit, from the statistical design point of view. The additional transistors have not degraded the robustness, since their contributions to the circuit performance are not too high anyway, thus, improvement of the circuit was possible for the same functional yield and standard deviation of the drain current mismatch.

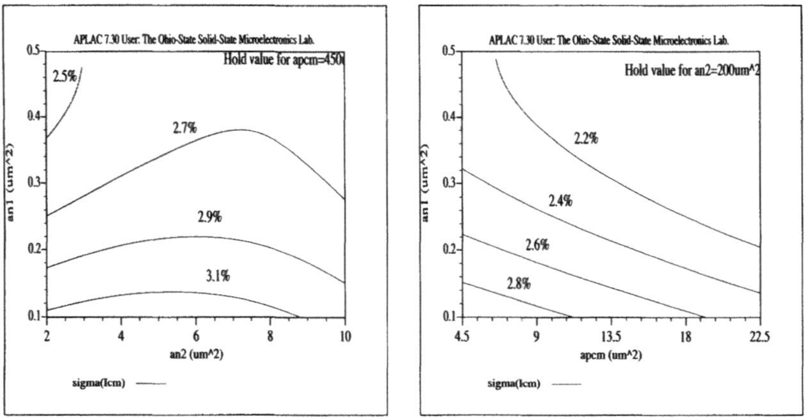

(a) (b)

Figure 7. Response surfaces for (a) a_{n1} vs. a_{n2} ($a_{pcm} = 450\ \mu\text{m}^2$), (b) a_{n1} vs. a_{pcm} ($a_{n2} = 200\ \mu\text{m}^2$)

It is possible to make a similar analysis of the results, as in the previous section. A specific standard deviation can be targeted, thus, a yield specification can be determined and the minimum area values achieving the targeted standard deviation and yield can be found; e.g., let us assume the goal of optimization is to obtain the minimum device area while achieving $I_{cm} \langle 6\%$, with a functional yield of 95%, or equivalently, to achieve the standard deviation of the relative drain current mismatch of 3% , since 95% is approximately. From Fig 7(a), the closest standard deviation values to 3.1% is 3%, thus, the response surface with this value will be considered. The minimum point on the response surface corresponding to 3% is found to be $a_{n1} = 11\,\mu\text{m}^2$, $a_{n2} = 200\,\mu\text{m}^2$ and $a_{pcm} = 450\,\mu\text{m}^2$. From the definition of W and L given by equation (11) and equation (12), $(W/L)_{n1} = 5.2/2.1$, $(W/L)_{n2} = 100/2$, and $(W/L)_{p1} = 150/3$. Thus , when these aspect ratios are used for transistors M_{n1} , M_{n2} and M_{pcm} , the standard deviation will not exceed 3%, and the functional yield will be 95% . Again , the only way to prove this result is to fabricate the circuit in large numbers and calculate the standard deviations from the measurement data. This way of evaluation also explains how to enchance the yield. By archieving optimization of the circuit, the tolerance of the circuit will be reduced, thus, when this specification is met, the yield will be achieved for a tighter limit, hence, the yield will be enhanced.

5. ROBUST LOW VOLTAGE OPAMP DESIGN

The persistent pursuit or low cost and high degree of integration by the IC industry has led to great advance in CMOS technology, most notably of which is the constant reduction in the minimum feature size of MOS transistors. The small feature size meets the requirement for low power and high speed in digital design, but provides serious challenges to analog design due to increased parameter variation. The traditional analog design involves an iterative process from design to manufacturing and then back to design for improvement. This process is time consuming and cost inefficient and is further complicated by the greater parameter variation associated with the modern CMOS process. Therefore, design for manufacturability becomes an important issue facing analog designers. The influence of modern VLSI process on analog design is best demonstrated in the design of the operational amplifier (opamp), one of the most fundamental building blocks and most challenging designers in analog signal processing systems. In recent low voltage ($\leq$ 3V) opamps have been the research focus of many industrial and academic institutes [2-4, 6, 21-25]. At The Ohio State University, Lin and Sakurai each developed unique constant g_m topologies to solve the problem of g_m variation seen in most low voltage opamps. These new topologies add complexity to circuitry, and with increased random fabrication variations could degrade the statistical performance of the opamp. One of the most critical performances of the opamp is the DC offset voltage. The DC offset is mostly random and directly affected by device parameter variation and circuit structure. Since random variations do not scale down with feature size supply voltage, DC offset would constitute a serious drawback of low voltage opamps and severely hinders its application in high precision circuits.

With the supply voltage of the digital IC reduced to 3.3V and lower,

Rail-to-rail input and output operation becomes a necessary feature for low voltage opamps. To process a rail-to-rail input, a low voltage CMOS opamp usually features complementary differential pairs as the input stage. The PMOS pair operates in the low common mode voltage range, while the NMOS pair is in the high common mode voltage range. Thus, three common mode operation regions exist for the input stage: PMOS only, PMOS + NMOS, and NMOS only. Without special circuit design,

These regions have varied g_m and, thus, g_m dependent characteristics, such as unity gain frequency, phase margin and DC gain will vary as common mode voltage changes. This g_m variations generates distortion and complicates phase compensation. Therefore, constant g_m is a very desirable feature for low voltage rail-to-rail opamps.

The opamps in this study were originally developed by Lin and Sakurai, respectively, and contain novel robust structures to maintain constant g_m in the input stage [22,23]. The robust design methodology employed in both designs eliminates the need for matching of the NMOS and PMOS input pairs. Although they share similar characteristics, the two opamps differ in the ways of keeping the input g_m constant. Sakurai's circuit consists of a monitor of a common mode current of the PMOS pair and a bias circuit, which adjusts the bias currents to the NMOS pairs to satisfy $\sqrt{K_p I_p} + \sqrt{K_n I_n} = constant$. Lin's design (Fig.8 (a)) contains a transconductance equalizer bias circuit and a maximum current selecting circuit. The bias circuit generates two constant currents to the PMOS pair in and the NMOS pair, and the two currents satisfy $\sqrt{K_p I_p} + \sqrt{K_n I_n}$ order to make g_m of the PMOS and NMOS pair equal, under their respective strong inversion regions. The maximum current selecting circuits switch the larger output current of the pairs to the following cascode gain stage and, thus, at any common mode voltage only one pair contributes to voltage amplification. The g_m variation in the middle common mode region is, therefore, avoided. Also, because pairs at their respective strong inversion regions have equal g_m, the of the input stage is, thus constant.

5.1 Robust Low Voltage Rail-to-Rail Opamp Architecture

Fig. 9(a) shows a robust, universal design strategy of low voltage rail-to-

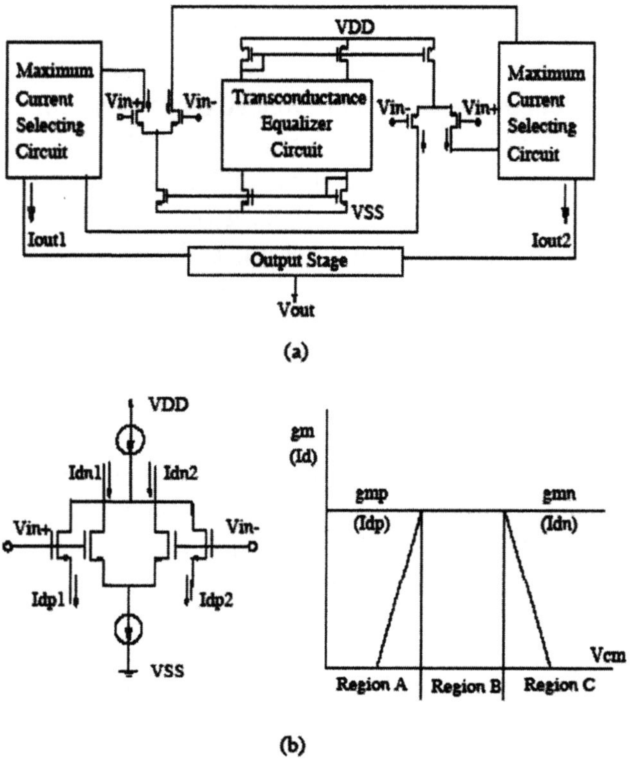

Figure 8. A robust design methodology for low voltage rail-to-rail opamps

rail opamps. The universality here refers to the fact that the strategy is technology independent and it can be implemented in any VLSI technology (Bipolar, GaAs, CMOS, etc.) with complementary devices. It also refers to the fact that in CMOS, the strategy is valid in both weak and strong inversion [4,6,21,22]. The maximum current selecting circuits direct the maximum current to the outputs I_{out1} and I_{out2} , and with the assumption thet these maximum currents are constant and equal, a constant g_m ia achieved (see Fig. 9(b)). The transconductance equalizer bias circuit circumvents the requirement of $K_p = K_n$ of the complementary differential pairs. Such a requirement is common in most rail-to-rail opamps reported prior to the works of Sakurai and Lin. The output stage in Fig. 6 with a class AB contro in used to subtract the currents and amplify the differential voltage and, if needed, to provide a buffer.

5.1.1 Basic Circuit Blocks

5.1.1.1 Complementary Differential Pairs and Maximum Current Selecting Circuit

The transconductance , g_m , is dominated by the tail current of a transistor. That is

$$g_m = \frac{dI_d}{dV_{gs}} = \sqrt{2K I_{bias}} \tag{15}$$

where $K = \frac{1}{2}\mu_{n,p} Cox \frac{W_{n,p}}{L_{n,p}}$ is the electron mobility, Cox is the gate oxide capacitance, and $\frac{W_{n,p}}{L_{n,p}}$ is the asoect ratio of the transistor. In the differential

$$I_{n1,n2} = \frac{I_{nbias}}{2} \pm g_{mn}\frac{v_{id}}{2} \tag{16}$$

$$I_{p1,p2} = \frac{I_{pbias}}{2} \pm g_{mp}\frac{v_{id}}{2} \tag{17}$$

pairs, the total instantaneous output currents , or the drain currents, can be expressed as [4,21,22]:
for NMOS differential pair, and PNMOS differential pairs. The currents, I_{pbias} and I_{nbias} , are tail currents of the PNMS pairs. v_{id} is the differential mode input voltage.

To maintain the tranconductance of the p-channel differential pairs equally at full swing range of common mode input voltage, we should have.

$$g_{mp} = \sqrt{2K_p I_{pbias}} = \sqrt{2K_n I_{nbias}} = g_{mn} = g_{m,max} \tag{18}$$

This means that $I_{pbias} = I_{pbias} = I_{pbias,max}$ should be selected from rail-to-rail, and $K_p = K_n$ should be assumed as the required condition.

Using the maximum current selecting circuits [4, 21, 22] shown in Fig. 9, the drain current pairs, (I_{n1}, I_{n2}) or (I_{p1}, I_{p2}) of the differential pairs are selected strainght at any common mode swing (see Fig. 8). Therefor, the output currents of the input stage the maximum current selecting circuit (I_{out1}, I_{out2}) , are equal to the maximum currents of (I_{p1}, I_{n2}). Obviously,the signal current can be extracted and expressed as

Consequently, the relationship in equation (18) guarantees that the

$$i_{out} = I_{out1} - I_{out2} = g_{m,max} \frac{v_{id}}{2} \qquad (19)$$

transconductance of the input stage is always kept at a constant value, $g_{m,max}$,in the entire common input voltage range.

5.1.1.2 Transconductance Equalizer Bias Circuit

The universal approach requires $K_p = K_n$. The fabrication process of integrated circuits could result in K_p ,K_n mismatch of the transistors. This mismatch always exists in fabricated circuits and hence, the performance of the opamp would not be as good as predicted. Recently [23] the process parameter variations, $K_p = K_n$. The fabrication process of integrated circuits could result in K_p, K_n mismatch of the transistors. This mismatch always exists in fabricated circuits and hence, the performance of the opamp would not be as good as predicted. Recently [23], the process parameter variations , μ_p , μ_n, in the 2 μm Orbit, 2 μm VTI, etc., used by the MOSIS service are studied. The deviations of μ_p , μ_n, in these processes are found to be almost 30%. This deviations may result in a large g_m variation in low voltage rail-to-rail opamps.

A tranconductance equalizer circuit [23] is employed to secure the constant transconductance even when $K_p \neq K_n$, resulting in a robust low Voltage opamp design. Let us assume that the given ratio of the electron mobility, xx , is approximately 2.8. If the design is carried out with this assumption, the value of the ratio will be between 2.2 and 2.3, with about 20% deviation. This deviation will cause the transconductance to vary around 10%. Under the conditions $K_p = K_n$ and $I_{bias} = I_{nbias}$ the total transconductance is kept constant. A more general robust approach should be based on the following condition to achieve equation (20):

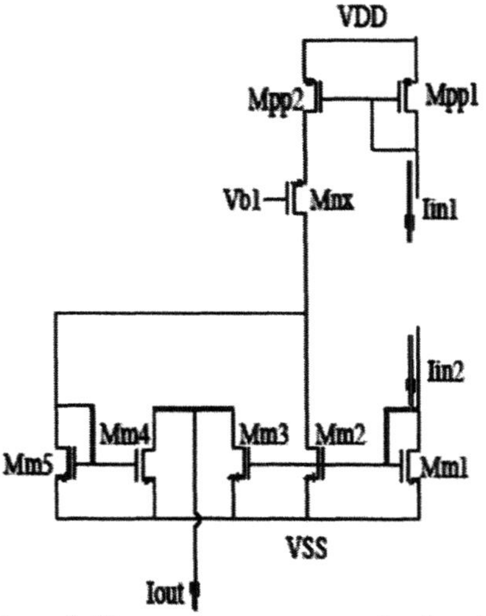

Figure 9. The maximum current selecting circuit

$$\sqrt{2K_p I_{pbias}} = \sqrt{2K_n I_{nbias}} \tag{20}$$

and a transconductance equalizer circuit is shown in Fig. 10. Note that the circuit is designed to operate only in strong inversion. The bias circuit is supplied by a constant bias current $5/4I_{nbias}$. The main part of the bias circuit is made up by M_{a1}, M_{a2}, M_{a3} and M_{a4}. Using KVL around the four transistors, we have

$$V_{sg2} + V_{gs3} = V_{sg1} + V_{gs4} \tag{21}$$

Because of the square-law characteristic and the two current mirrors, M_{a5}, M_{a6}, M_{a8}, M_{a9}, with a current ratio of 4:1, and , $V_{gs} = V_t + \sqrt{\frac{2I_d}{K}}$ equation (21) can be expressed as

$$\sqrt{2K_p I_{pbias}} = \sqrt{2K_n I_{nbias}} \tag{22}$$

where the pair of transistors, M_{a1}, M_{a2}, (M_{a3}, M_{a4}) have the same threshold voltages, V_{tp}, (V_{tn},). Obviously, the equation means that g_{mn} is equal to g_{mp}. As K_p is not equal to $K_{n,}$, the bias circuit provides two different bias current. I_{nbias},I_{nbias} , to two differential pairs for archieving $g_{mn(max)}$=$g_{mp(max)}$. Acording to the above equation, no matter how the ratio of K_p and K_n is changed, g_{mn} and g_{mp} should be equal with the adjustment of the bias currents and regardless of the ratio of K_p and K_n. Of course, the simple current mirror transistors (M_{a5}, M_{a6}, M_{a7}, M_{a8}, M_{a9}, M_{a10}) may suffer from channel length modulation. This non-ideal effect limits the compensation capability of the transconductance equalizer circuit within a narrow range of bias currents and limited ratios of K_p and K_n..

$$\frac{I_{nbias}}{I_{pbias}} = \frac{K_p(1 \pm x)}{K_n(1 \pm x)} \tag{23}$$

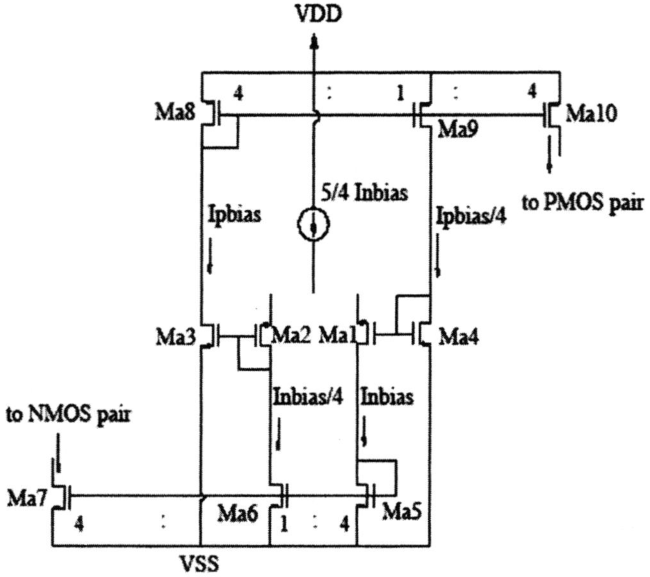

Figure 10. The transconductance equalizer bias circuit

6. A SINGLE STAGE OPAMP DESIGN EXAMPLE

Using the basic circuit blocks already discussed, Lin's input stage is implemented as shown in Fig. 6.11. The simulation results in Fig. 6.12 show that the operations of the input stage with the different bias currents, 25μA, 20μA, 15μA have different constant g_m. With $K_p \neq K_n$, this input stage is simulated in the bias current range from 10μA to 40μA. Fig. 6.12(b) indicates that the drain currents of the input differential pairs with bias currents , 25μA, 20μA, 15μA compensate for the mismatch of p-channel and n-channel input differentail pairs. The simulated variation percentage of g_m is between 4.61% and 6.67%.

One of the main goals of this opamp design is to reduce the variation of transconductance due to mismatch of the p-channel and n-channel differential pairs. Some simulations based on the assumption of the ratio variations of K_p and K_n are done and shown in Fig. 6.13. the deviation relationship of bias currents, and K_p and K_n is defined as

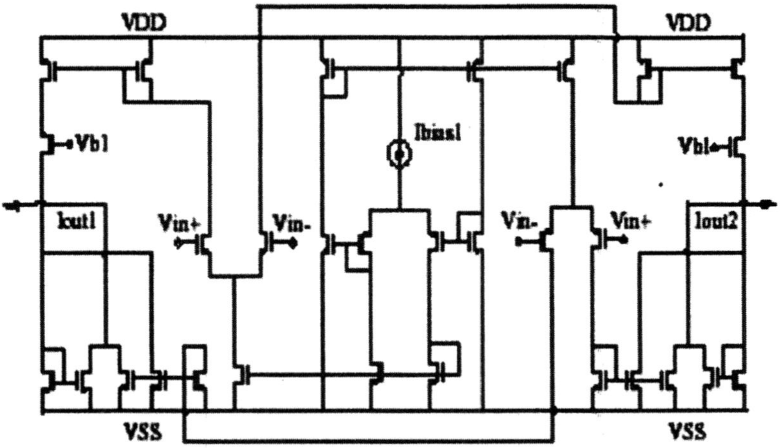

Figure 11. The robust rail-to-rail constant g_m input stage

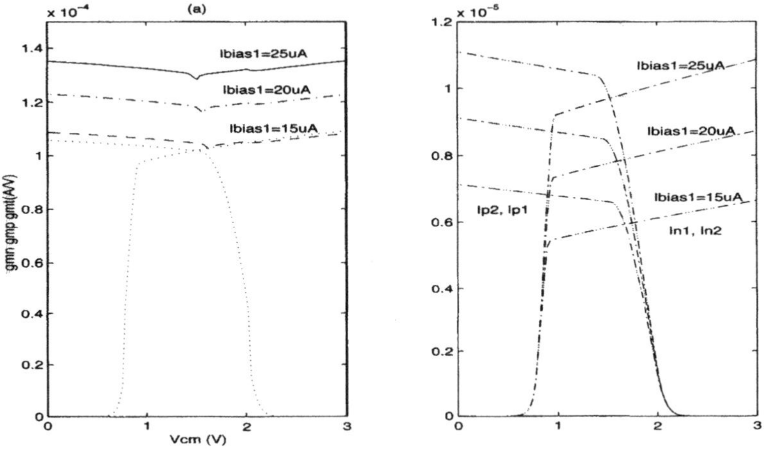

Figure 12. (a) The transconductance of the input stage with different bias currents, 25μA, 20μA, 15μA and (b) The drain currents of the input stage with 15μA, 20μA, 25μA bias currents compensate for the mismatch of p-channel and n-channel input differential pairs

where x is a random number between 1% and 15%.

The smallest and largest variation with 2%, 5%, 10% K_p and K_n deviations are displayed in Fig. 6.13. Obviously, the transconductance of the nput stage is still expected to be a stable constant value. The range of the estimated g_m variation percentage with ±15% K_p and K_n deviations id from 4.76% to 5.14% for the smallest variation case and from 5.92% to 18% for the largest variation case. The major g_m variation (Fig. 6.13(b)) is caused by the reason that the output currents (I_{p1}, I_{p2}) enter into the trioderegion around V_{cm}=1.5V. From the simulation of frequency response of the input stage, the dominant pole is located around 11MHz. Fig. 14 shows that two other different MOSIS models (N550, N71V) are used to test the function of the transconductance equalizer circuit, the results are similar to the mismatch estimation of K_p and K_n .

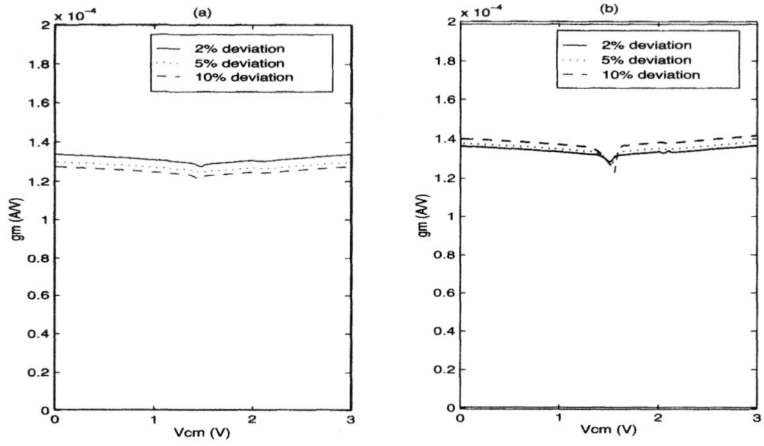

Figure 13. The simulated (a) smallest and (b) largest gm variation cases of 2%, 5% and 10 Kp and Kn deviation

In order to obtain the information of total transconductance due to current subtraction (equation (19)) and obtain voltage amplification, an output stage for this purpose is necessary for this opamp. The output stage [24] is composed of a cascodegain and class AB control circuits, which is power efficient and has a rail-to-rail output swing. The output currents, I_{out1} and I_{out2} are respectively, connected with the output stage. A complete two stage single-ended operational amplifier is shown in Fig. 15.

The simulation is based on the SPICE level-2 model with 3V power suply. Neither M_{o1} nor M_{o2} is turned off; the output stage always works from rail-to-rail. The minimum standby currents help to increase the speed during the operation. The input-output voltage characteristic of the unity gain configuration is shown in Fig. 16(a). The output voltage almost matches with the input voltage, except for a range of 0.006V within both upper and lower rails. The open loop frequency and phase response of the opamp is shown in figure 16(b). Because of the constant transconductance of the input stage, the low frequency open loop gain, A_O, and the unity gain frequency, f_U, are constant from rail-to-rail.

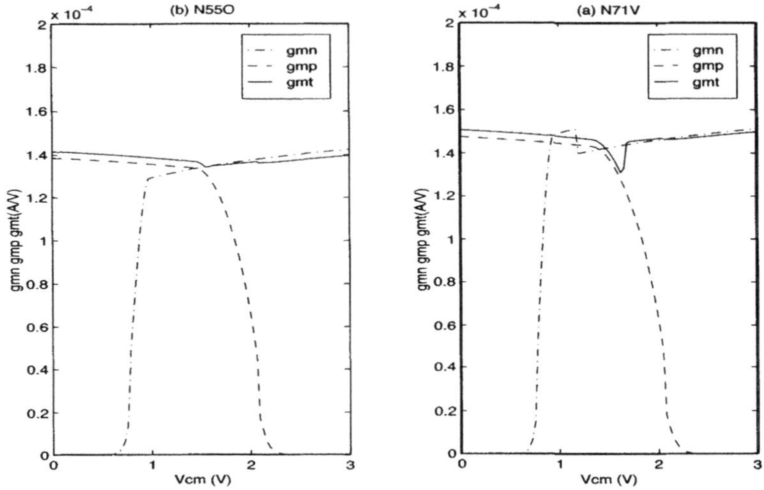

Figure 14. The two different MODIS 2μm models (a) N550, (b) N71V are used to test the function of the transconductance equalizer circuit

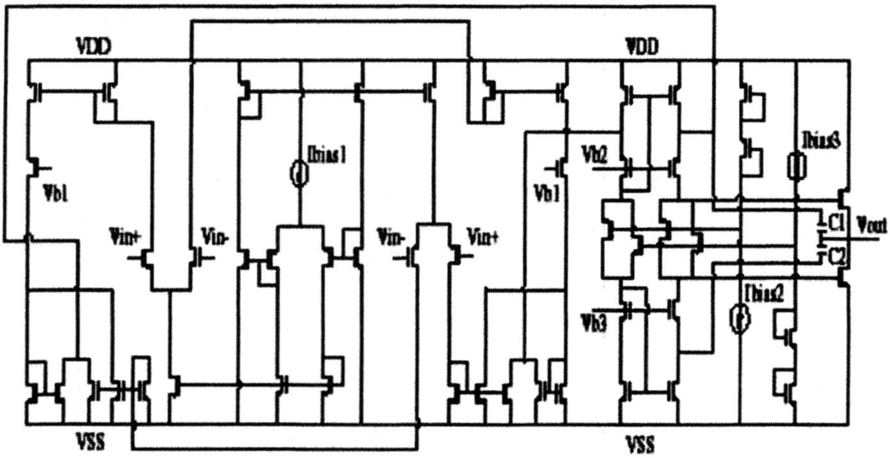

Figure 15. A two stage single-ended operational amplifier

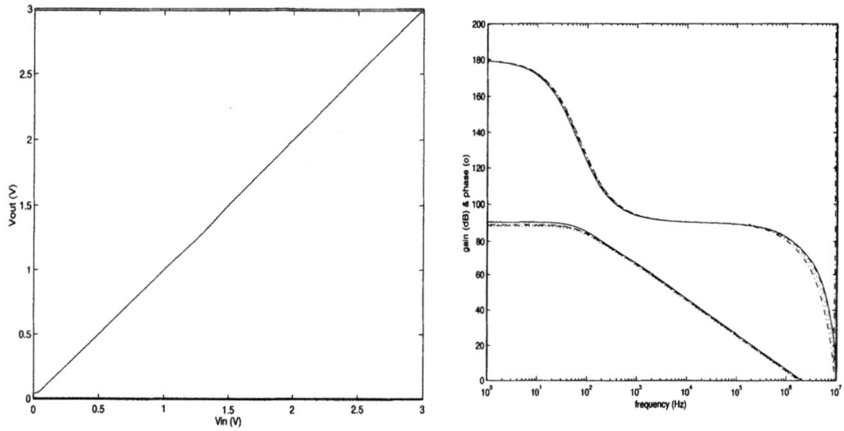

(a) (b)

Figure 16. (a) The input-output voltage characteristics of the unity gain configuration, (b) Open loop frequency response of the opamp

7. A TWO STAGE LOW VOLTAGE OPAMP DESIGN

The performance of the opamp with I_{bias}= 25μA and R_L= 10km c_L=20p is specified in Table 9.

Table 9. The performance of the opamp with with I_{bias} = 25μA and R_L = 10km c_L =20p

Parameter	Value
DC input offset voltage	-1328μV
Commom mode input voltage	OV to 3V
Output voltage swing	0.006 to 2.95V
Open loop gain	86dB
Unity gain frequency	1.246MHz
Phase margin	$\geq 76^0$
Slew rate ↑	1.01V/μs
Slew rate ↓	1.33V/μs
CMRR @ 1MHz	79dB
Positive PSRR @ DC	112dB
Negative PSRR @ DC	92dB
Total power dissipation	0.60mW

The DC input offset voltage is 1.3μV. The swing range of the output voltage is between 0.06V and 2.95V without distortion. The low frequency open loop gain approaches 86dB. The unity gain frequency is 1.24MHz, and the phase margin is always larger than 76degrees. Table 10 shows the total harmonic distortion (THD) with a 10kHz sinusoidal input signal wave.

Table 10. THD of the opamp as the unity gain buffer

V*in* (V)	THD (%)
0.1	0.0038
0.2	0.0062
0.3	0.0078
0.4	0.0080
0.5	0.0075
0.6	0.0051
0.7	0.0057
0.8	0.0065
0.9	0.0105
0.10	0.0174
0.11	0.0236
0.12	0.0293
0.13	0.0311
0.14	0.0246
0.15	0.3777

The total harmonic distortion is less than 0.04% while the amplitude of the input wave within 1.4V is applied to the opamp. The THD at 1kHz indicates a very low distortion percentage which is 0.0004755%, 0.0004833%, 0.0005842% for V+*cm* = 0.5V, 1.5V, 2.5, respectively. The low voltage CMOS rail-to-rail opamp is simulated by SPICE and APLAC [16]. The chip has been fabricated using 2μm ORBIT process by MOSIS. The total area of the opamp is 563 X 650μm^2.

8. STATISTICAL DESIGN AND OPTIMIZATION OF LOW VOLTAGE OPAMPS

Sakurai's [23] and Lin's [25] designs are two representative constant g_m schemes seen is several low voltage rail-to-rail opamps. However, their DC offset characteristics have not been analyzed and, thus, it is necessary to perform statistical simulations on both circuits and optimize the layouts to obtain better offset performance [26]. It is also interesting to compare the offset of the two circuits, which share many nominal characteristics. To facilitate the study of the offset, the one-stage structure was chosen. The reasons for this choice are;

1. Most on chip applications of opamps do not require the output stage
2. Due to the high DC gain (70dB) of the one-stage opamp omitting the output stage does not affect the conclusion on the offset characteristics as 0.3V offset voltge in the output stage only causes negligible 0.1mV offset at the input
3. The one-stage configuration saves computational time.

The structure of the circuit used in this study are adoped from the original circuits with some changes to reduce chip area and DC offset. The modifications were aimed at optimized DC offset performance and were justified by the statistical simulations.

Opamp1 is a one-stage version of the circuit developed by Lin with the folowing modifications:

- The current mirror ratio in the transconductance equalizer bias circuit is reduced to 1:2 from 1:4. This reduction halves the areas of M_{bn4}, M_{ns}, M_{bp3} and M_{ps} without adverse effects. With this modification the bias current I_{bias1} need to be changed to 30μA to maintain the tail curent of the differential pairs.
- Large W/L ratios of transistors in the maximum current selection circuit are reduced to save the chip area and to accommodate the levels for statistical runs. Again this change does not have any significant effect on the circuit performance.
- The two NMOS pairs of the cascode gain stage in the original circuit is redundant to part of the maximum current selecting circuit (M_{an3}, M_{an4}, M_{con1}, M_{xn3}, M_{xn4} and M_{con2}) and thus can be spared. This change not only simplifies the circuit but also reduces offset due to less matching errors.

Opamp2 has an identical structure to Sakurai's opamp1 with the following modifications to reduce the chip area and facilitate comparison with opamp1:

- The width of M_8 and M_p are halved to reduce the tail current to the same level as in opamp1 (about 20μA).

- The areas of transistors in the signal path are reduced to meet the level requirement of statistical runs
- The bias of the cascode stage is decreased to $I_{DS,MC3} = 8\mu A$ to meet the DC gain requirement.

The modified one-stage opamps are shown in the figure 17 and figure 18, respectively.

After modification, both amplifiers have similar g_m and maximum bias current in the input stage and their DC performance and frequency response are also comparable. The simulated characteristics of both opamps with the transistor areas set to the lowest possible values (equal to level -1 values in the statistical experiments) are listed in table 11.

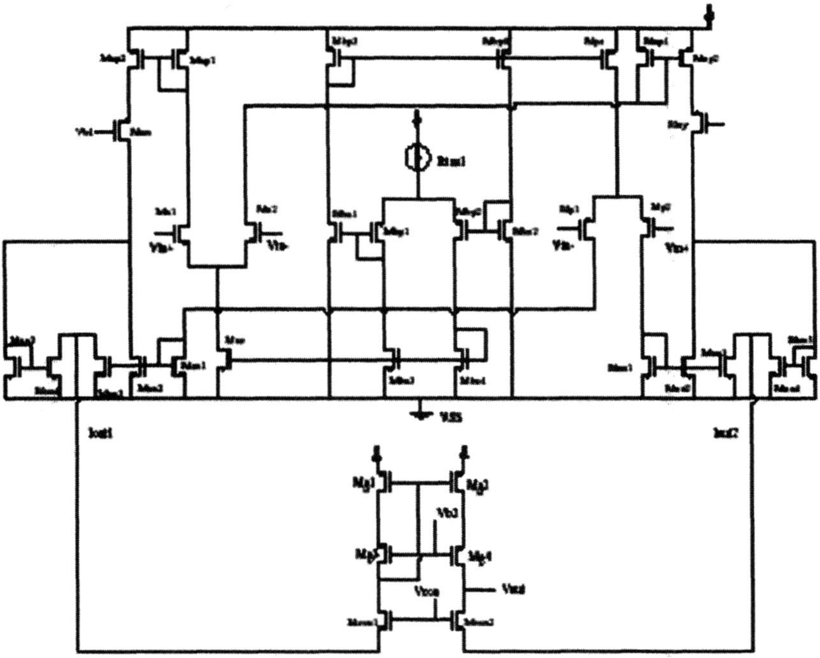

Figure17. The structure of opamp1.

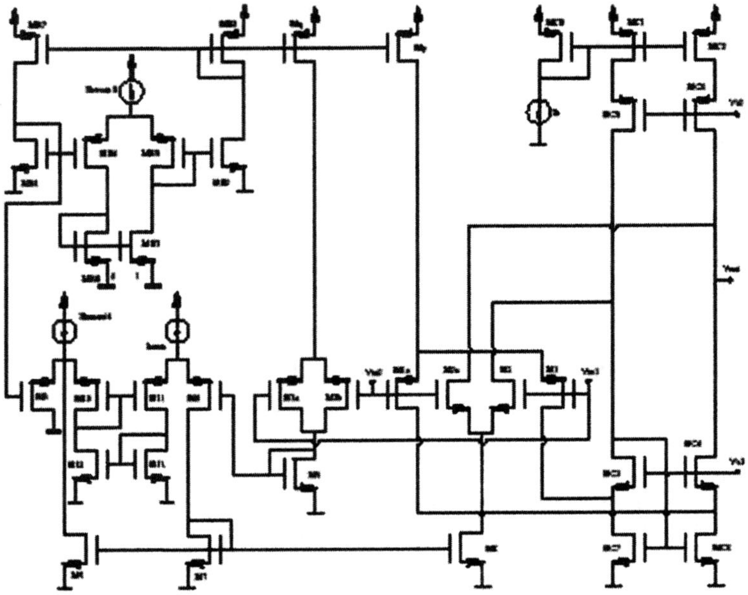

Figure 18. The structure of opamp2.

Table 11. Simulated characteristics of opamp 1 and opamp 2

Opamps	Systematic offset	DC gain	Unity gain frequency	Phase margin
Opamp1	-52μV	73dB	2.3MHz	83°
Opamp2	-87μV	69dB	2.5MHz	83°

8.1 DC Offset Simulation

The randomness of the offset is simulated by calling the SMOS model in a Monte-Carlo loop, and the standard deviation of offset voltage calculated from those of all the Monte-Carlo runs.

To apply the SMOS model to the opamps, additional layout information must be known. That means that the location of each transistor on the chip must be specified. For opamp1, the XY coordinates of transistors are extracted from the actual layout of Lin's opamp. Simulation with the SMOS model on the original layout revealed an unacceptably high DC offset with σ_{voff} = 9mV. This value is so high that in order to achieve a yield of 95%, a chip with DC offset of 18mV (2σ) has to be accepted. An inspection of the layout revealed that common-centroid or interdigitized structure was only applied to the input differential pairs and the cascode stage. However, simple circuit analysis suggests that matching in the maximum current selection circuit may have a significant effect on the offset. Therefore interdigitized or common-centroid layout was applied to this transistor pairs later, and the DC offset voltage dropped to 5.9mV. Further analysis of the circuit showed that the two NMOS pairs in the cascode gain stage are redundant and removing them should not have adverse impact on the circuit operation, but only further reduces the offset voltage as the matching error of these transistors no longer contribute to the total offset. This was again proved by the simulation with σ_{voff} now going down to 5.1mV. Thus the usefulness of the SMOS model in the layout design was successfully demonstrated.

The actual layout of opamp2 is not available and the XY coodinate of each transistor is assumed based on the transistor size and the circuit structure. Again all current mirrors and differential pairs in the signal path are assumed to be common-centroid to reduce the offset. This opamp has a DC offset of σ_{voff}=2.1mV.

Thought the offset performance of both opamps is acceptable for many applications, such as switch-C filter, sample-hold amplifier, optimization is possible to bring the DC offset voltage further down without significantly increasing the schip area and without compromising other performances.

8.2 Statistical Experiments

In IC circuits, the device mismatch is a function of the device area and the separation distance, but not the aspect ratio, which mainly affects an nominal circuit performances. Thus, the DC offset of an opamp is mainly affected by the device area and the separation distance [11-13]. With all the input differential pairs and current mirrors laid out interdigitized or common-centroid in both opamps, the effect of the separation distance cannot be further minimized. Therefore, the efforts were focused on optimizing device areas. Moreover, circuit analysis and simulations show that the transistor areas of the bias circuit only influence the common mode performance, but not differential mode characteristics such as DC offset. Thus, to simplify the experimental design, only the areas of the transistors on the signal path were selected as the input variables. Specifically in opamp1, tge areas of M_{n1} - M_{n2} - M_{p1} - M_{p2}, M_{ap1} - M_{ap2} - M_{xp1} - M_{xp2}, $M_{an1\text{-}5}$ - $M_{xn1\text{-}5}$, M_{n1} - M_{nx} - M_{ny}, M_{con1} - M_{con2}, M_{g1} - M_{g2} and M_{g3} - M_{g4}; and in opamp2, those of $M_{n1\text{-}1a}$, $M_{p1\text{-}1a}$, $M_{c1\text{-}2}$, $M_{c3\text{-}4}$, $M_{c5\text{-}6}$, $M_{c7\text{-}8}$, were used as input variables for the statistical experiments.

During the experiment the change in transistor area will slighly affect nominal circuit performance. Thus, to derive an empirical model of the DC offset for a functional opamp, the aspect ratios need to be adjusted to center them to the desired specifications. This design centering can be done using the automatic optimization method Minmax of APLAC [16]. Minmax adjusts the optimization variables to minimize the maximum error between actual simulation results and desired goals. For the ease of optimization, only the DC gain (70 dB) was specified as the optimization goal and, thus, the systematic offset is low as a consequence of the hight DC gain. The attempt to include AC characteristics (unity gain frequency, phase margin) into the optimization loop proved to be difficult with Minmax. However, the AC performance does not affect the study of the offset and can always be adjusted manually by trimming the compensation capacitor.

As discussed previously, a two-step experimental procedure can be adopted in this study. The first step is the two-level Placket-Bueman screening test to eliminate the less influential variables. The area and level of each transistor is specified in table 12 and table 13.

Table 12. Area level and assingnments for each transistor in opamp1

Transistors	M_{n1-2} M_{p1-2}	M_{ap1-2} M_{xp1-2}	M_{nx} M_{ny}	M_{an1-5} M_{xn1-5}	M_{con1} M_{con2}	M_{g1} M_{g2}	M_{g3} M_{g4}
Area Sysmbol	A_n	a_{ap}	a_{nx}	a_{an}	a_{con}	a_{g1}	a_{g3}
(-1) (μm^2)	200	100	200	200	200	200	200
(+1) (μm^2)	2000	1000	2000	2000	2000	2000	2000

Table 13. Area level and assingnments for each transistor opamp2

Transistors	M_{2-2a}	M_{C1-2}	M_{C3-4}	M_{C5-6}	M_{C7-8}	M_{1-1a}
Area Sysmbol	A_n	A_{c1}	A_{c3}	A_{c5}	A_{c7}	A_p
(-1) (μm^2)	500	500	500	500	500	500
(+1) (μm^2)	2500	2500	2500	2500	2500	2500

Since there are seven and six variables for opamp1 and opamp2, respectively, eight runs for each circuit are needed. Table 14 and table 15 show the design matrices and the simulation results. Each run in the design matrices of the Mont-Carlo loop for n = 500 and takes about 40 seconds of computation time on a HP 715/100 workstation.

Table 14. Placket-Burman design matrix and results for opamp1

Run	a_n	a_{ap}	a_{nx}	a_{an}	a_{con}	a_{g1}	a_{g3}	σV_{off}(mV)
1	1	-1	-1	1	-1	1	1	4.46
2	1	1	-1	-1	1	-1	1	3.34
3	1	1	1	-1	-1	1	-1	2.67
4	-1	1	1	1	-1	-1	1	2.53
5	1	-1	1	1	1	-1	-1	5.16
6	-1	1	-1	1	1	1	-1	2.44
7	-1	-1	1	-1	1	1	1	5.05
8	-1	-1	-1	-1	-1	-1	-1	5.15

Table 15. Placket-Burman design matrix and results for opamp2

Run	a_n	a_{c1}	a_{c3}	a_{c5}	a_{c7}	a_p	σV_{off}(mV)
1	1	-1	-1	1	-1	1	2.02
2	1	1	-1	-1	1	-1	1.51
3	1	1	1	-1	-1	1	1.56
4	-1	1	1	1	-1	-1	1.87
5	1	-1	1	1	1	-1	2.04
6	-1	1	-1	1	1	1	0.99
7	-1	-1	1	-1	1	1	1.74
8	-1	-1	-1	-1	-1	-1	2.24

The SS values for each transistor in the opamp are given in table 16 and table 17, respectively. In opamp1, the areas of M_{ap1} - M_{ap2} - M_{xp1} - M_{xp2} and $M_{an1\text{-}5}$ - $M_{xn1\text{-}5}$, and in opamp2, the areas of $M_{p1\text{-}1a}$ - $M_{c1\text{-}2}$ and $M_{c7\text{-}8}$ are the most influential variables constituting more than 95% of their respective total SS. Thus, they were carried over to the following model building experiments. The other areas were set to level -1 during the following experiments and optimization due to their insignificant effects.

Table 16. SS values and contribution of each transistor in opamp1

Factors	SS	%
a_n	0.03	0.3
a_p	9.76	92.2
a_{nx}	0	0
a_{an}	0.32	3.0
a_{con}	0.18	1.7
a_{g1}	0.31	2.9
a_{g3}	0	0

Table 17. SS values and contribution of each transistor in opamp2

Factors	SS	%
A_n	0.011	1.0
a_{c1}	0.560	52.0
a_{c3}	0.025	2.3
a_{c5}	0.002	0.2
a_{c7}	0.251	23.3
A_p	0.228	21.2

The second step of the statistical experiment is the three-level Box-Behnken model building experiment. The design matrices are shown in table 18 and table 19 and the simulation results are also listed. Nine and fifteen runs are required for opamp1 and opamp2, respectively, and a Monte-Carlo loop of n = 1000 were included in each run.

Table 18. Box-Behnken design matrix and results for opamp1

Run	a_{ap}	a_n	σ_{Voff}(mV)
1	-1	-1	5.513
2	0	-1	3.525
3	1	-1	3.442
4	-1	0	5.259
5	0	0	2.880
6	0	1	2.788
7	1	-1	5.156
8	1	0	2.760
9	1	1	2.617

Following the simulation regression models were constructed using the statistical tool Minilab [19]. The standard deviation of DC offset (mV) of opamp1 is expressed as follows:

$$\sigma_{Voff} = 6.484 - 0.814a_1 - 0.0726a_2 + 0.053a_1^{\,2} \qquad (24)$$

where $a_1 = a_{ap}/100$ and $a_2 = a_{an}/100$. With the R^2 =99,8%, the model for the DC offset of opamp1 fits the experimental results very well. The σ_{Voff} and a_{ap} and a_{an} relationship is also illustrated in figure 19. As is obvious from the plot, the offset decreaseis almost monotonically as a_1 or a_2 increases, and a_1 is dominant in affecting offset.

In opamp2, σ_{Voff} is fitted with the following quadratic model:

$$\sigma_{Voff} = 2.9613 - 0.7479a_1 - 0.4057a_2 + 0.3641a_3 + 0.2508a_1^{\,2} + 0.0739a_2^{\,2} + 0.0684a_3^{\,2} - 0.7479a_1 a_3 \qquad (25)$$

where $a_1 = a_{ap}/1000$, $a_2 = a_{C1}/1000$ and $a_3 = a_{C7}/1000$. The a_1a_2 and a_2a_3 terms are omitted because their t values are between 0.5 and 0.5. The R^2 is 88.2% and the fitness is acceptable considering the number of variables involved.

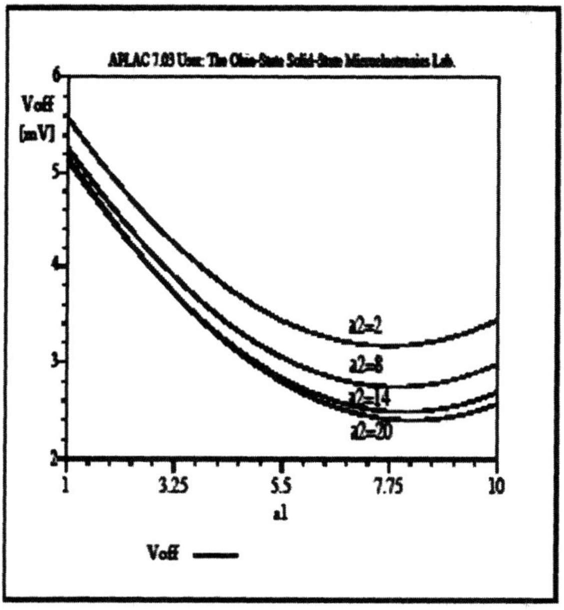

Figure19. Standard deviation of DC offset vs. device areas for opamp1

8.3 Optimization

Since both empirical models have been constructed, it becomes a mathematical problem to optimize the DC offset. The DC offset is a function of the chip area utilization producing less offset. Thus, efforts need to be made to meet the specified offset requirement with the minimum chip area. The mathematical approach for this study can be simple exhaustive search because both models are quadratic functions with only two or three variables.

The optimization on opamp1 is shown graphically to illustrate the exhaustive search method. The combinationof a_1 and a_2 that gives standard deviations of 2.5mV, 3.0mV, 3.5mV and 4.0 mV are solved from equation 24 and plotted in figure 20.

The area utilization (100 μ^2) of transistors $M_{ap1\text{-}2}$ - $M_{xp1\text{-}2}$ - $M_{an1\text{-}5}$ and $M_{xn1\text{-}5}$ is expressed as $Area = 4a_1 + 10\ a_2$, and is shown in figure 21. a_1 is limited to [1,10] and a_2 is limited to [2,20] because the model was characterized in these ranges and the opamp may not function with smaller areas. .The discontinuity of curves with $\sigma_{Voff} = 3.0mV$ in both figures is due to the complex roots of equation 24.

The conditions that give the minimum area utilization are listed listed in table 20. The simulated DC offset with optimal a_1 and a_2 combination are also listed. It is obvious that the simulation results are in good agreement with the desired DC offsets. This close matching proves the accuracy of the empirical model.

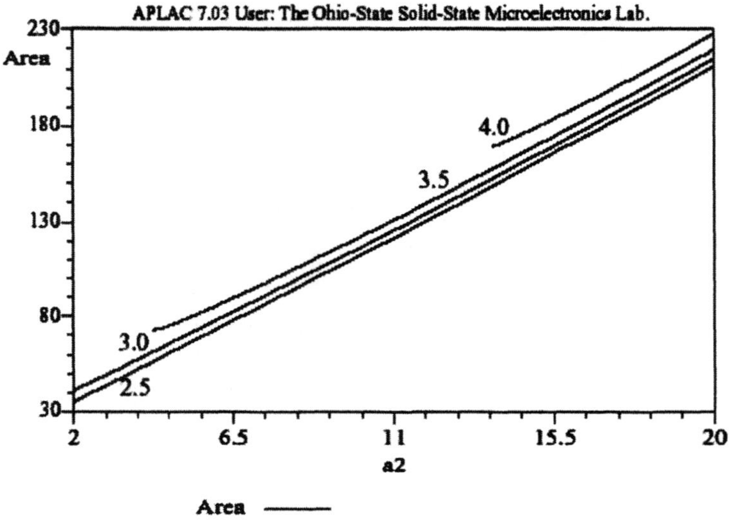

Figure 20. The contour curves of the standard deviation of DC offset (mV)

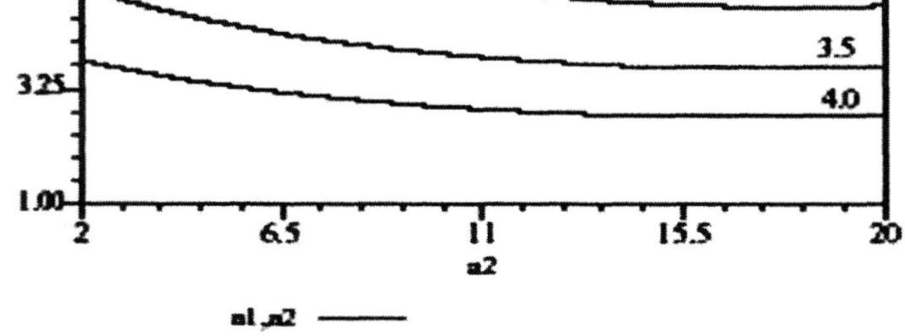

Figure 21. The area utilization (100μm) vs a_2 under various requirements of DC offset standard deviation (mV)

The empirical model for opamp2 consists of three variables and thus the graohical method is not suitable. Instead, a program routine was written in APLAC to do the exhaustive search. This routine scans two of the variables from the lower bound (0.5) to the upper bound (2.5) independently. The third variable is solved based on equation25 for a given σ_{Voff}, and the area utilization $2a_1 + 2\ a_2 + 2\ a_3$ is calculated for each scanning point. The a_1, a_2 and a_3 values with the minimum area utilization can be found by searching all the calculation results. For σ_{Voff} requirement of 1.5mV, this method gives the optimal device areas as $a_1 = 1267$, $a_2 = 1120$ and $a_3 = 1594$. With this combination, the simulated DC offset is 1.47mV.

8.4 Comparison and Discussion

The SMOS model coupled with the statistical tools proved to be very effective in the DC offset optimization of low voltage rail-to-rail constant g_m CMOS opamps. The accuracy of this method was verified by circuit simulation. Now with the empirical models of DC offset established and verified, the comparison between the two topologies can be made.

With the common mode voltage sweeping rail-to-rail, opamp1 has nearly constant current in the cascode gain stage because of the maximum current selecting circuit. Thus, the DC gain of opamp1 is nearly constant across the common mode range. Opamp2 relies on the dynamic bias to maintain the constant g_m and, thus, the current through the gain stage varies. This variation alters DC gain and larger device areas have to be assigned to the transistors to keep the gain comparable to that of opamp1.

The offset of opampl is heavily dependent on the device areas of the maximum current selecting circuit, as expected. Because of this dependency, effects of other devices are masked. Opamp2 has lower offset voltage with the same area utilization because of the absence of extra current mirrors in the signal path. Without the masking effects of these current mirrors, both input differential pairs and the cascode stage contribute to the total offset in opamp2. The two common-gate pairs, $M_{c3\text{-}4}$ and $M_{c5\text{-}6}$, in the gain stage function to enhance the output impedance and, thus, are not involved in current manipulation. Therefore, they do not significantly affect the total offset. The bias circuit in opamp2 adjusts the tail current of the NMOS input differential pair by monitoring the tail current of the PMOS pair. At zero common mode voltage, the PMOS pair is still in the strong inversion region and, thus, the tail current and g_m are close to the maximum. With the regulation of the constant g_m bias circuit, little current and g_m are allowed for the NMOS pair. Thus, the contribution of NMOS pair mismatch to the offset is negligible. With the same reasoning, at a high common mode voltage, the NMOS input pair will dominate in offset generation.

9. CONCLUSION

This chapter introduced basic design techniques for low voltage analog MOS integrated circuits at both the schematic as well as the physical layout levels. The chapter demonstrates the critical need to perform statistical design and optimization in order to enhance both the functional yield and reliability of low voltage low power analog VLSI circuits, and produce cost effective manufacturable designs.

10. ACKNOWLEDGMENTS

The support of both SRC and NSF to both low voltage and low power design and statistical modeling activities at The Ohio State University are appreciated. Financial support from TUBITAK (The Scientific and Technical Research Council of Thrkey) is gratefully acknowledged.

11. REFERENCES

[1] Meindl, J.D., "Low Power Microelectronics: Retrospect Prospect", Proceedings of the IEEE, vol.83, No.4, April 1995

[2] Huijsing, J.H. and Linbarger, D., "Low Voltage Operational Ampli fier With Rail-to-Rail input and output ranges", IEEE Journal of Solid-State Circuits, vol.SC-20, pp.1144-1150. December 1985

[3] Hogervorst, R., Wiegerink, R.J., de Jong, P.A.L., Fonderi, L., Wassenaar, R.F., and Huijsing, J.H., "CMOS Low Voltage Operational Amplifier with Constant gm Rail-to-Rail Input Stage", Proceedings of the IEEE International Symposium on Circuits and Systems, pp. 2876-2879,1992

[4] Hwang, C., Motamed, A., and Ismail, M., "Universal Constant- g_m Input Stage Architecture for Low Voltage Opamps", IEEE Transactions on Circuits and Systems, vol.42, pp.2876-2879,1992

[5] Abel, C., Michael, C., Ismail, M., Teng, C.S., and Lahri, R., "Reliability Characterization of Mixed-Signal Chips", IEEE Circuits and Devices Magazine, pp.8-10, July 1997

[6] Motamed, A., "Low-voltage Analog VLSI Circuits and Signal Processing", Ph.D. Dissertation in Electrical Engineering, The Ohic State University, 1996

[7] Hwang, C., Hyogo, A., Ismail, M. and Kim, H., "LV CMOS Analog VLSI Composite Cell Design and Its Application to High Speed Multiplier", IEEJ ,st Int'l Analog VLSI Workshop, ECT-97-48, pp95-98, The Ohio State University, USA, May 1997

[8] Seevinck, E. and Wassenaar, R. F., "A Versatile CMOS Lineai Transconductor / Square-law Function Circuit", IEEE Journal oi Solid State Circuits, SC-22, pp366-377, June 1987

[9] Hyogo, A., Hwang, C., Ismail, and M. Sekine, K., "LV/LP CMOS Square-Law Composite Transistors for Analog VLSI Applications" IEEJ 1st International Analog VLSI Workshop, ECT-97-59, pp139-143, The Ohio State University, USA, May 1997

[10] Motamed, A., Hwang, C., and Ismail, M., "CMOS Exponentia Current-to-Voltage Converter", Electronics Letters, Vol.33, No.12., pp998-1000, June 5th, 1997

[11] Tarim, T.B. and Ismail, M., "Statistical Design and Yield Enhancement of CMOS Analog VLSI Circuits", IEEE Circuits and Device' Magazine, pp.12-22, March 1999
[12] C. Michael and M. Ismail, Statistical Modeling for Computer-Aided Design of MOS VLSI Circuits, Kluwer Academic Publishers, Boston. 1993
[13] Pelgrom, M.J.M., Duinmaiger, A.C.J., and Welbers, A.P.G.. "Matching Properties of MOS Transistors", IEEE Journal of Solid-State Circuits, vol. SC-24, pp.1433-1439, October 1989
[14] Shyn, J.-B., Temes, G.C., and Krummenacher, F., "Random Erroi Effects in Matched MOS Capacitors and Current Sources", IEEE Journal of Solid State Circuits, vol.SC-19, pp.948-955, December 1984
[15] Lakshmikumar, K.R., Hadaway, R.A., and Copeland, M.A., "Characterization and Modeling of Mismatch in MOS Transistors for Precision Analog Design",IEEE Journal of Solid State Circuits, vol.SC-21, pp.1057-1066, December 1986
[16] Helsinki University of Technology, Circuit Theory Laboratory and Nokia Research Center, APLAC-An Object Oriented Analog Circuit Simulator and Design Tool, 7.1 User's Manual and Reference Manual, 1997
[17] G.E.P. Box, Empirical Model Building and Response Surfaces, John Wiley&Son, 1987
[18] D.C. Montgomery, Design and Analysis of Experiments, New York:
Wiley, 1997
[19] Minitab, Statistical Software, Release 12, User's Manual, 1997, http://www.minitab.com
[20] Tarim, T.B., Statistical Design and Yield Enhancement of Low Voltage CMOS Analog VLSI Circuits, Ph.D. dissertation, Electronics and Communications Engineering Department, Istanbul Technical University, Istanbul, Turkey, 1999
[21] Constant Transconductance Design Methodology and Implementations thereof Inventors: A. Motamed, C.Hwang, and M. Ismail, Filed: August 14,1995, Issued: Feb. 3,1998, U.S., Patent No. 5,714,906
[22] Hwang, C., Motamed, A., and Ismail, M., "Theory and Design of Universal Low Voltage OpAmps", Chapter 1.1.3 in Emerging Technologies: Designing Low Power Digital Systems, R. Cavin and W. Lin (Editors), ISCAS 1996 Thtorial Book, May 1996, IEEE Catalog No. 96TH8189
[23] Sakurai, S. and Ismail, M., "Robust Design of Rail-to-Rail CMOS Operational Amplifiers for a Low Power Supply Voltage", IEEE Journal of Solid State Circuits, vol.31, no.2, pp.146-156, February 1996
[24] Hogervorst, R., Tero, J.P., Eschauzier, G.H. and Huijsing, J.H., "A Compact Power Efficient 3V CMOS Rail-to-Rail Input/Output Operational Amplifier for VLSI Cell Libraries", IEEE Journal of Solid State Circuits, pp.1505-1513, December 1994
[25] Lin, C.-H., Chi, H., Hwang, C., and Ismail, M., "A Robust Low Voltage CMOS Rail-to-Rail OpAmp Architecture", Proceedings of the 1st Analog VLSI Workshop, the IEEE of Japan, pp.17-22, Columbus, OH, May 1997
[26] Clii, H.," Statistical Design and Optimization of Low Voltage Railto-Rail CMOS OpAmps" M.S. Thesis, The Ohio State University, May, 1997

12. BIOGRAPHY

Tuna B. Tarim received her B.Sc., M.Sc. and Ph.D. degrees in Electronics Engineering in 1992, 1994 and 1999, respectively, from Istanbul Technical University, Electrical and Electronics Engineering Faculty,

Electronics and Communication Engineering Department, Istanbul, Turkey. She was a visiting scholar with the Analog VLSI Lab at The Ohio State University, Electrical Engineering Department from June 1997 to December 1999. She is currently a Wireless Design Engineer with the Mixed Signal Wireless Division at Texas Instruments, Inc., Dallas, Texas, in the Customer Owned Tooling (COT) group. Her current research interests include functional yield enhancement, statistical design and optimization of VLSI circuits, analog/mixed-signal VLSI and device modeling. She published over 25 papers in these areas. She is a member of IEEE. She has served in the technical committees of many conferences. Currently she is the tutorial chair of the International Symposium of Quality Electronic Design. She is the associate editor of the Mixed Signal Letter section of the International Journal of Analog Integrated Circuits and Signal Processing. She serves as chair of the Analog Signal Processing Technical Chapter of the IEEE CAS society.

Chi-Hung Lin was born in Talchung, Taiwan, in 1970. He received the M.S. degree in electrical engineering from The Ohio State University, Columbus, U.S.A., in 1997. He is currently a Ph.D. candidate at The Ohio State University and is a co-operative designer in the wireless IC design center, IBM, USA. His research interests include high-speed data converters, filters and RF CMOS circuits. He is currently engaged in the design of high-speed Sigma-Delta data converters foi next generation communications.

Mohammed Ismail (S'80-M'82-SM'84-F'97) received the B.S. and M.S. degrees in electronics and telecommunications engineering from Calro University in 1974 and 1978, and the Ph.D. in electrical engineering from the University of Manitoba in 1983. He is a professor with the Department of Electrical Engineering, The Ohio State University and is the founder and director of the Analog VLSI Lab. He has held several positions previously in both industry and academia and has served as a corporate consultant to nearly 30 companies in the United States and abroad. He held visiting appointments at the Norwegian Institute of Technology University of Oslo, University of Twente, Tokyo Institute of Technology, Helsinki University of Technology and the Swedish Royal Institute of Technology. He has authored many publications on VLSI circuit design and signal processing and has been awarded several patents in the area of analog VLSI. He has co-edited and co-authored several books including a text on Analog VLSI Signal and Information Processing, (McCraw Hill, 1994). He advised the work of 14 Ph.D. students, 42 M.S. students, and 16 visiting scholars. His current interests include low-voltage/low-power VLSI circuits RF and mixed signal VLSI circuits for wireless communications, statistical computer-aided design and optimization, integrated circuits for image, video and multimedia applications and VLSI information processing systems. He gives intensive courses to industry in these areas. Dr. Ismail has been the recipient of several awards including the IEEE 1984 Outstanding Teacher Award, the NSF Presidential Young Investigator Award in 1985, the OSU Lumley Research Award in 1993 and 1997, the SRC Inventor Recognition Awards in 1992 and 1993, and a Fulbright/Nokia fellowship Award in 1995. He is the founder of the International Journal of Analog Integrated Circuits and Signal Processing and serves as the Journal's Editor-In-Chief (North America). He has served the IEEE in many editorial and administrative capacities, including General Chair of the 29th Midwest Symposium CAS, member of the CAS Society Board of Governors, chair of the CAS Analog Signal Processing Technical Committee, the Circuits and Systems (CAS) Society's editor of the IEEE Circuits and Devices Magazine, founder and co-editor of The Chip, a column in the magazine, and associate editor of the IEEE Transactions on Circuits And Systems, IEEE Transactions on Neural Networks, IEEE Transactions on VLSI Systems and IEEE Transactions on Multimedia. He cofounded Micrys, Inc. (formerly ChipWorks, Inc.), a commercial VLSI design company specialized in analog and mixed-signal ASIC's. Dr.Ismail is a Fellow of IEEE.

Chapter 7

Ultralow-Voltage Memory Circuits

Kiyoo Itoh
Central Research Laboratory, Hitachi, Ltd.,Kokubunji, Tokyo 185, Japan, Tel. +81-423-23-1111, Fax +81-423-27-7699, E-mail k-itoh@crl.hitachi.co.jp

Key words: DRAM, SRAM, high S/N cell, subthreshold current, gate-source backbias, multi-VT, dynamic VT, SOI

Abstract: The key design issues for ultralow-voltage (0.5-2 V) memory circuits are reviewed in terms of stable memory-cell operation, subthreshold current reduction, suppression of or compensation for design-parameter variations, and a single power supply and its standardization. The results obtained are as follows. (1) In DRAMs, coupled with high signal-to-noise-ratio memory-cell designs, the gate-source offset driving schemes suppress the cell-transistor subthreshold current increased by reduction of the threshold-voltage (VT). The gate-source self-backbiasing scheme drastically reduces the subthreshold current of the peripheral circuit, especially of iterative circuit blocks. Multi-VT and dynamic VT schemes recently proposed for logic LSI chips are also effective in reducing the subthreshold current. Various on-chip voltage generators and converters are becoming increasingly important in suppressing or compensating for the design parameter variations and in implementing and standardizing a single power supply. In SRAMs, a boosted power-supply scheme for the cell will eventually become necessary in order to accommodate the cell transistor's high-VT needed to suppress a huge array subthreshold current. (2) SOI circuits are attractive in terms of ultralow-voltage operation although the floating body issue remains unsolved. Intrinsic fluctuations of FET parameters caused by random microscopic fluctuations in dopant atoms in an extremely short channel of 0.1 mm or so may limit ultralow-voltage operation, thus requiring new device designs.

1. INTRODUCTION

The low-power RAM circuit is a major area of interest in low-power LSI research. Successive advances in low-power RAM-circuits have been able to suppress chip-power consumption which increases with increasing memory capacity, chip area and speed. As a result, coupled with high-density memory-cell technology, these advances have allowed chip power consumption to be maintained or lowered [1-5], although the memory capacity of DRAM chips has rapidly increased by 6 orders (1Kb to 1Gb) over the last 25 years. Low-power RAM circuits are also essential to meet the increasingly high throughput requirements of personal computers (PCs) [3]. Moreover, they form the basis not only of other LSI memory chips such as flash and ROM, but they are also the basis of on-chip memory subsystems such as embedded DRAMs (merged DRAM and logic) [3] and SRAM caches that have both become increasingly important in modern memory systems. The designers of the low-power RAM circuits developed so far have focused on three key issues [1]: reducing in charging capacitance, operating voltage, and dc current. Of these issues a reduction in the operating voltage has become relatively important not only to reduce power, but also to ensure device reliability in scaled-down devices, and also to extend the use of LSIs to battery-based portable systems. Historically, DRAM researchers initiated and then led the field in low-voltage LSI research, because their first priority was on higher density chips which were obtained through scaled-down FETs consequently resulting in lower breakdown voltages. Low-voltage (2 to 3 V) circuits have been used in actual 16-Mb and 64-Mb DRAM products, although their external supply voltages are 5 V or 3.3 V, being internally lowered by on-chip voltage-down converters to standardize power-supply [1,2]. Recent exploratory research on ultralow-voltage operations of 1 V or less suggests the great potential of CMOS circuits, although reducing voltage inevitably imposes memory-cell development focusing attention on a high signal-to-noise ratio (S/N) design [1-5].

In this chapter the key design issues affecting ultralow-voltage (0.5 V - 2 V) RAMs are reviewed in terms of memory-cell and peripheral circuit designs. First, these issues are summarized. Second, modern DRAM designs are discussed which focus on high S/N memory-cell design, and on different types of state-of-the-art ultralow-voltage peripheral circuits to suppress the subthreshold current. Third, SRAM designs which emphasize cell driving schemes are described. Finally, the potential of SOI technology and unacceptably large design-parameter fluctuations expected in 0.1 μm LSI era are discussed.

2. DESIGN ISSUES FOR ULTRALOW-VOLTAGE RAMS

The key design issues affecting ultralow-voltage RAMs can be summarized as stable memory-cell operation, subthreshold current reduction, suppression of or compensation for device parameter variations, and single power-supply and its standardization.

2.1 Stable-Memory Cell Operation

The memory cell distinguishes most large memory-capacitor RAM chips (Figure 1) from MPU/ASIC chips. The RAM chips peripheral circuit which excludes the memory array has almost the same circuit configuration as that of a MPU/ASIC chip. Reducing power-supply voltage, VDD, inevitably decreases the signal charge (Qs) of the memory cell, as shown in Figure 2, causing small cell-signal voltage on the data line. Cell operation, therefore, is susceptible to various noise sources, resulting in unstable operation [2]. Hence, an increase in Qs is critical to extend the lower limit of VDD.

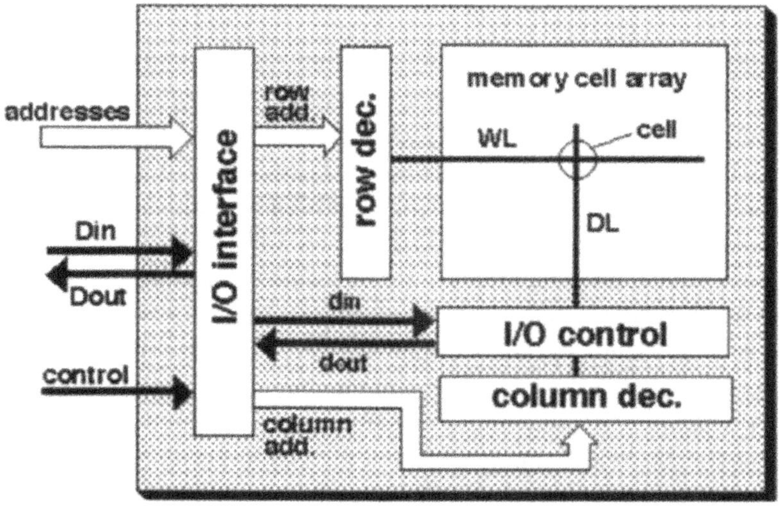

Figure 1. Memory chip configuration

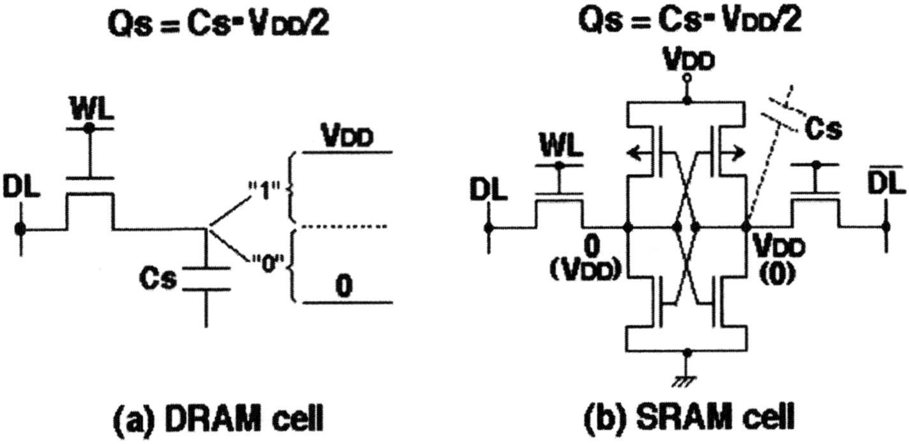

Figure 2. Cell signal charge (Qs) vs. supply voltage (VDD).

2.2 Subthreshold Current Reduction

MOSFET threshold voltage (VT) scaling has been highlighted as an emerging issue [1] in the simultaneous achievement of low-voltage and high-speed operation. The high-speed operation of CMOS circuits necessitates scaled-down VT because speed is roughly inversely proportional to VDD-VT. However, the MOSFET subthreshold dc current starts to increase exponentially with decreasing VT [4].

To evaluate subthreshold current caused by VT scaling, the definition of MOSFET threshold voltage must be clarified. There are two kinds of VT: extrapolated VT and constant-current VT, as shown in Figure 3. The extrapolated VT is defined by extrapolating the saturation current on the √IDS-VGS plane, and neglecting the tailing current actually developed at approximately VT. Our major concern is the subthreshold current which is developed at 0 V VGS. If VT is high enough, the subthreshold current is 0. With decreasing VT, however, substantial subthreshold current starts to develop at a VT higher than expected. This current is not expressed in this definition, although circuit designers are familiar with extrapolated VT. Thus, constant current VT is indispensable in evaluating current. The VT is defined as a VGS for a given current density on the log IDS-VGS plane. This constant current VT is empirically estimated to be smaller than the extrapolated VT by about 0.2 V for a current density of 2nA/mm.

$$I_L = W \frac{I_0}{W_0} E l_0 - V_T / S \qquad (1)$$

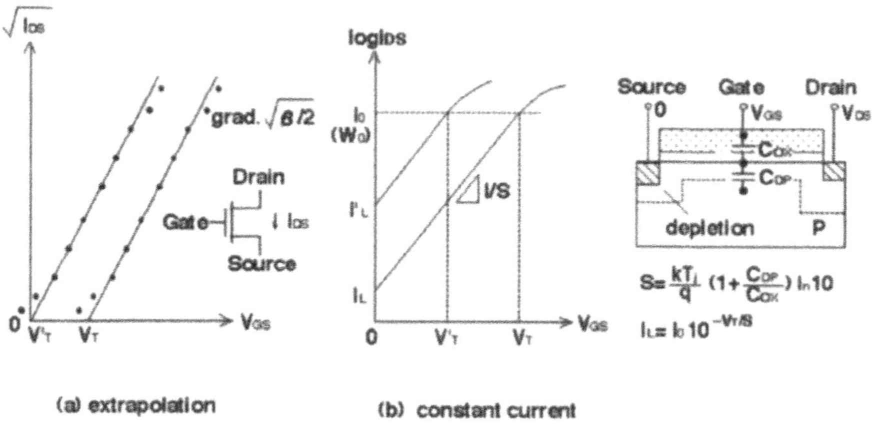

Figure 3. VT definition [4].

where W is the gate width of the FET, I0/W0 is the current density to define VT, and S is the subthreshold swing which is expressed by FET parameters and junction temperature (Tj), as shown in the figure.

The resultant dc current degrades the cell-characteristics, and eventually dominates chip current, as explained below, reducing the low-power advantage of CMOS circuits that we take for granted today. It also disables the detection of defective chips by monitoring the quiescent power-supply current (so-called IDDQ test). Therefore, reducing subthreshold current is essential not only in designing gigabit RAM chips of 0.1mm or less in the future, but also in designing ultralow-voltage megabit RAM chips using existing fabrication processes tailored to scaled VT.

1. **Memory-Cell Current:** The subthreshold leakage current of a DRAM cell FET flows from the cell storage node to the data line while the data line is at a low level, as shown in Figure 4. This degrades the data-retention time of DRAM cells. In SRAMs, two sources for leakage current are established in a cell. As a result of current accumulation in numerous cells a SRAM array suffers from huge data-retention current along with decreasing VT [6], as shown in Figure 5. Thus, in principle, a cell-transistor VT cannot be scaled down. In particular, of all LSIs the DRAM cell needs the highest VT to ensure prolonged data retention time,

as will be explained later. This V_T restriction makes the design of ultralow-voltage more difficult.

2. ***Peripheral-Circuit Current:*** At present, in the subthreshold current issue, attention is mainly being paid to the standby period, since the V_T is still too high. With a further reduction in V_T, however, even the numerous circuits, especially iterative circuit blocks that are inactive during the active period, will start to generate subthreshold current. They eventually dominate the active current of the chip, as shown in Figure 6 [1]. Here I_{DC} is the dc chip current caused by subthreshold current, and IAC is the charging current for capacitive loading, while IACT is the total active current of the chip, that is, the sum of the IDC and IAC. Note that the total channel width of the iterative circuit blocks is overwhelmingly large, and this increases as memory capacity increases. Thus the circuit blocks are responsible for the IDC increase. Even a DRAM chip as small as 16-Mb suffers from increased chip current when VT is reduced despite using existing fabrication processes. The following distinguishing features of a peripheral circuit on a RAM chip, however, make the reduction of subthreshold current easier than in a MPU/ASIC chip despite the similar circuit configurations.

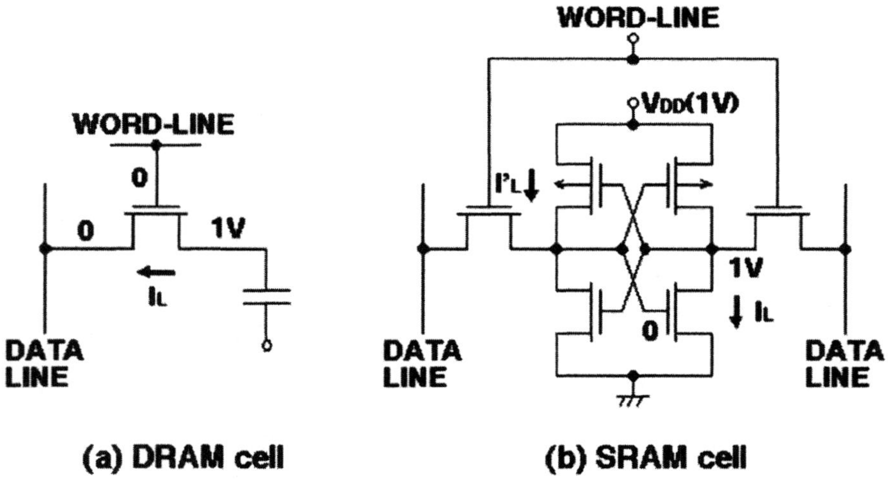

Figure 4. Degradation of memory cell characteristics due to V_T reduction.

4. **Slower Memory Cycle Time:** A physically large memory-cell array that occupies over 50% of the chip [5], a relatively large capacitance and a high resistance in word lines and data (or bit) lines [3], and a small memory-cell signal necessitating succeeding amplification are all responsible for a memory cycle time slower than that in a MPU/ASIC

chip. Moreover, the circuits are active only for a short period within the "long" memory cycle time, allowing an additional control time so as to reduce the subthreshold current.

5. **More Iterative Circuit Blocks:** A RAM chip incorporates many kinds of iterative circuit blocks such as a memory-cell array, row/column decoders and their relevant drivers, sense amplifiers, address buffers, and I/O buffers. Note that almost all circuits in each iterative circuit block are inactive even in the active period. This enables simple and effective control in reducing the subthreshold-current of each block.

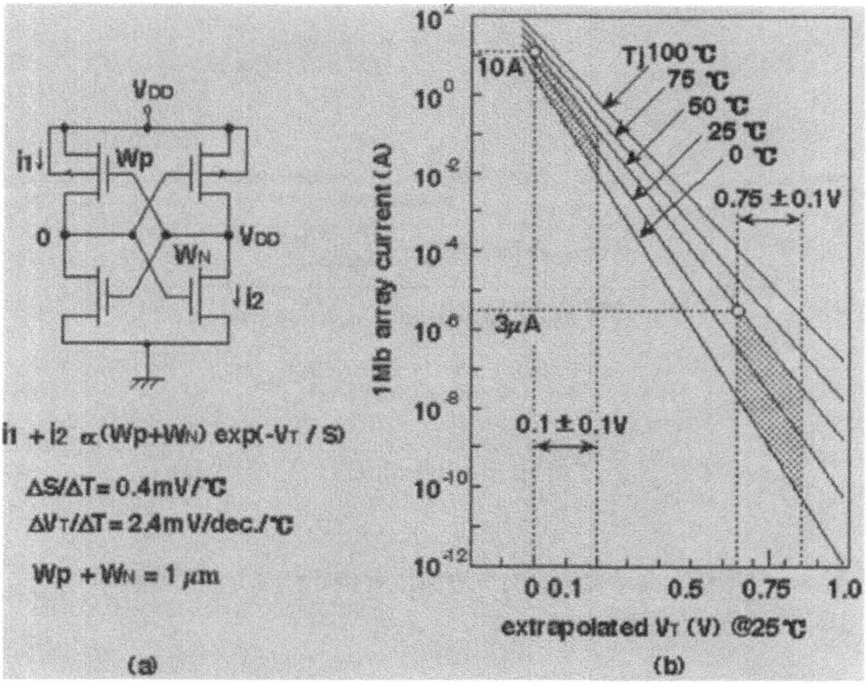

Figure 5. Calculated subthreshold current of a SRAM cell-array [6]. (a) The current sources in a full CMOS cell. (b) Subthreshold current of a 1Mb SRAM array versus the extrapolated VT.

6. **Incorporation of Input-Predetermined Logic:** Once a memory cycle starts to randomly select a memory cell using a few external clocks and address signals, the peripheral circuits do not work as random logic within the memory cycle. Thus, the designer can predict which FETs in the chip will cut off not only during the standby period, but also during

the active period. This also favors RAM chips with reduced subthreshold current.

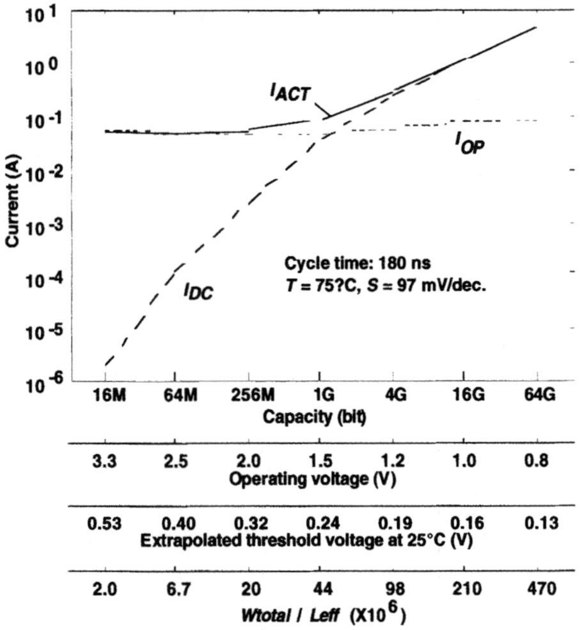

Figure 6. Estimated active current of DRAMs [1].

2.3 Suppression of or Compensation for Design-Parameter

2.3.1 Variations

Suppression or compensation circuits [6] for variations in design parameters such as VT, channel length, temperature, and VDD are very important. Device-parameter variations are unavoidably introduced during volume production. Even fixed variations increase chip-to-chip speed-variations with lowering VDD. Unfortunately, the miniaturization of FETs increase variations, causing unexpectedly large speed-variations at ultralow VDD. Unregulated battery power supply makes the design more complicated causing further speed-variations.

Figure 7 shows an example of speed variations assuming VT=0.15 V and L=0.1 mm for FETs of 0.6-mm channel length (L) and 15-nm gate oxide thickness (t_{OX}). Here memory capacity, chip area and all dimensions except those for FETs are assumed to be fixed. A voltage set of VDD=3 V and VT=0.45 V allows a normalized access-time spread of 0.7 to 1.5 according to two combinations of VT/L, that is, -0.15V/-0.1 mm and +0.15V/+0.1 mm. Another voltage set of VDD=1 V and VT=0.15 V, however, increases the spread from 1.6 to 5.6 with a nominal value of 3. Obviously, the VDD reduction not only degrades speed, but also the speed spread for fixed design-parameter variations. If a high performance FET of 0.3-mm L and 7.5-nm t_{OX} is used, almost the same nominal access time at 1-V VDD as at VDD=3 V could be obtained. The speed spread of 0.4 to 2.3 expanded by the same VT and L can be remarkably narrowed to 0.7 to 1.5, if both VT and L are scaled down to 0.5. Note that in addition to temperature increase, a VT decrease of about 0.1 V increases subthreshold current 10-fold. Thus suppressing or compensating for design-parameter variations is the key to achieving ultralow-voltage design.

2.4 Single Power-Supply and Power-Supply Standardization

General-purpose-use RAM chips have needed a single power-supply while keeping the power-supply standardization [2], in spite of the two power supplies of recent MPU/ASIC chips, exemplified by 3.3 V for the I/O and 2.5 V for the internal core circuit. In the course of lowering the operating voltage of general purpose RAMs,three-power-supply operation of 12 V (VDD), 5 V (Vcc only for the I/O interface circuit), and -5 V (VBB for substrate bias) changed to single power-supply operation of 5 V VDD in the 64-Kb generation, and then to 3.3 V in the 64-Mb generation. On-chip generators [5], as will be shown later, such as VBB generators, voltage-down converters, voltage-up converters, half-VDD generators, and reference-voltage generators have contributed to single VDD operation. The choice of a standard VDD continues to be one of the most important, serious, and urgent concerns because the power supply is always closely related to almost all design parameters. However, the recent excessive rapid down scaling of CMOS devices and strong demand of battery operation are making standardization difficult. In fact, VDD is still controversial in the 256-Mb and 1-Gb generations, although many attempts at low-voltage (2V to 0.5V) operation have been made.

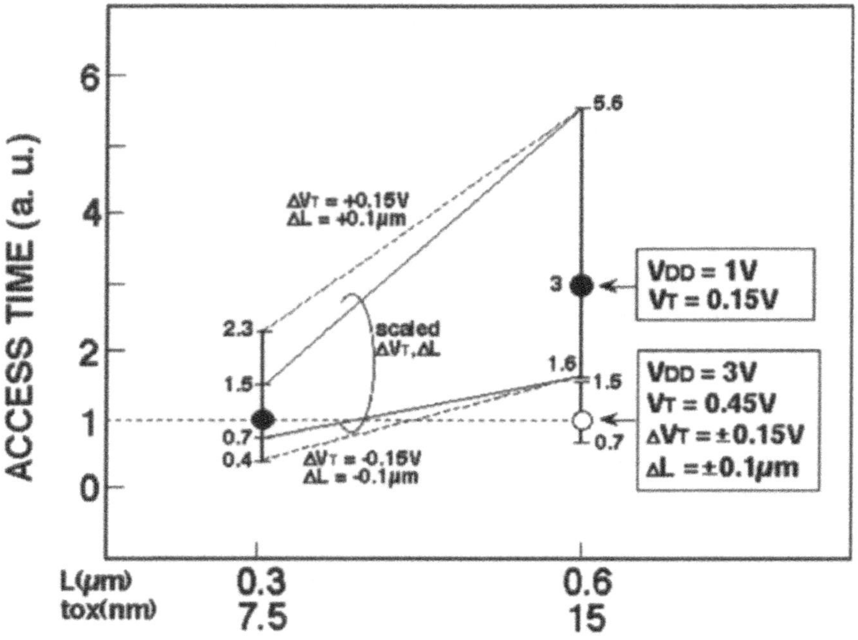

Figure 7. Speed variation for design-parameter variations. Memory capacity, chip area and all dimensions of devices except FETs are fixed.

3. DRAM CIRCUITS

Before going into further detail on the design issues described previously DRAM basic operations need to be explain ed. DRAM cell operation comprises read, write and refresh operations which are closely related to each other. In all operations data-line (DL) precharging and word-line (WL) activation (Figure 8) are common. They are done by equalizing all pairs of data lines to a floating voltage of a half-VDD by turning off the precharge circuit, and then activating a selected word line. In read operation, the stored data voltages, VDD or 0 V, of each cell along the word line are read out on the corresponding data line while using the other data line as a reference. As a result of charge sharing, the signal voltage (υ_S) developed on the floating data line is inherently small (100-200 mV) because the data-line parasitic capacitance (CD) is much larger than the cell storage-capacitance (Cs). Hence the original large signal component (VDD/2) at the storage-node collapses to υ_S. This destructive readout characteristic necessitates successive amplification and restoration operations for a selected cell on every data line. This is performed by a latch-type CMOS sense amplifier on each data line. Each sense amplifier, however, operates slowly, especially at ultra-low VDD. This is because of poor driving capability deriving from a half-VDD (the lowest voltage in the chip) operation, and high-density layout within a small data-line pair pitch.

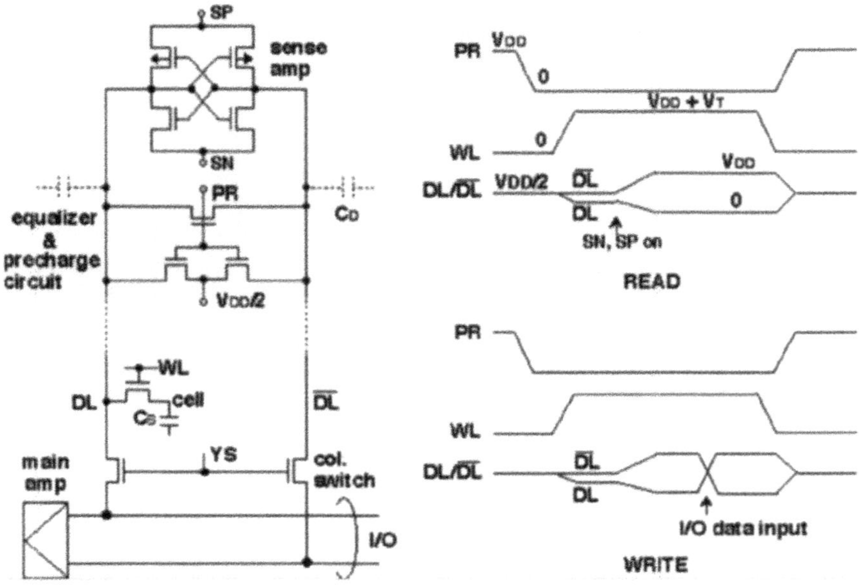

Figure 8. DRAM cell operation.

Thus, one major obstacle to ultralow operation is in the sense amplifier. Write operation is always accompanied by preceding read operation. After almost completing the above amplification, a set of differential data-in voltages of VDD and 0 V is inputted from I/O lines to the selected pair of data line. Hence the old cell data is replaced by the new. Note that the above read operation is simultaneously done for remaining cells on the selected word line to avoid the destruction of information. In refresh operation, the stored voltage of each cell degraded by the leakage current is restored by a refresh operation that is almost the same as for the read operation. This is done by reading the data of cells on the word line and restoring these for each word line so that all the cells retain the data for at least t_{REFmax}. Here t_{REFmax} is the maximum refresh time for the cell which is guaranteed in catalog specifications, exemplified by t_{REFmax}=64 ms for a 64-Mb chip. Thus, each cell is periodically refreshed at the interval of t_{REFmax}, although each has a data-retention time longer than t_{REFmax}.

3.1 Stable Memory - Cell Operation

3.1.1 Read Operation

To ensure successful sensing even for read operation just before refresh operation, the signal voltage must be larger than the noise voltage [2]. The signal voltage is expressed by

$$v_S = \frac{1}{C_D + C_S}\left(\frac{C_S A E_2^{VDD}}{2} + L\mathring{A} E_{RFE\,max} - Q_C\right) v_N \quad (2)$$

Here full write and full read operations are assumed, which are done by word bootstrapping VWL ≥ VDD + VT so that VT drop is eliminated.Hence the maximum data-line voltage of VDD is fully stored in the cell, and the stored voltage is fully utilized as a signal voltage on the data line. The formula above can be changed with the charge expression below

$$Qs > QL + Qc + QN \quad (3)$$

Where

Qs: signal charge (= CsVDD/2),

QL: leakage charge (= ILt_{REFmax}),

Qc: soft-error critical charge, that is, the maximum charge collected at the cell storage node by a-particle hittings,

QN: noise charge (~ CDuN if CD >> Cs).

This implies that the signal charge must exceed the total effective noise charge which is composed of leakage charge, soft-error critical charge and data-line noise charge, as shown in Figure 9. The relationship must be maintained despite VDD reduction through high S/N techniques aimed at larger Qs and less noise-charge components.

Larger Qs *[1-5,7]:* In addition to full write operation and a thinner capacitor-insulator, a half-VDD capacitor plate enabling double capacitance, vertical capacitors such as stacked and trench capacitors, and a capacitor over data (or bit) line (COB) structure affording an increased capacitor area have all been especially important in obtaining a larger Qs in products [1]. Figure 10 shows the trend in the memory-cell structure [7]. As a result, Qs has been able to be maintained despite reductions in both cell area and VDD, as shown in Figure 11 [1]. Ultralow-voltage operation calls for further developments in extremely high permittivity materials [3] to obtain a larger Cs.

Less Q_L: Fortunately, V_{DD} reduction reduces Q_L because of less p-n junction leakage current due to less stress voltage. The subthreshold current issue that is to be discussed later emerges instead.

Less Q_C: Qc also tends to reduce with a reduction in V_{DD}.

***Less Q_N** [1-5,7]:* A reduction in Q_N has been achieved by reducing of C_D and υ_N. C_D was reduced by using a multi-divided data-line structure combined with a shared I/O (Figure 12). υ_N was reduced by using folded data-line arrangement and a twisted data-line structure.

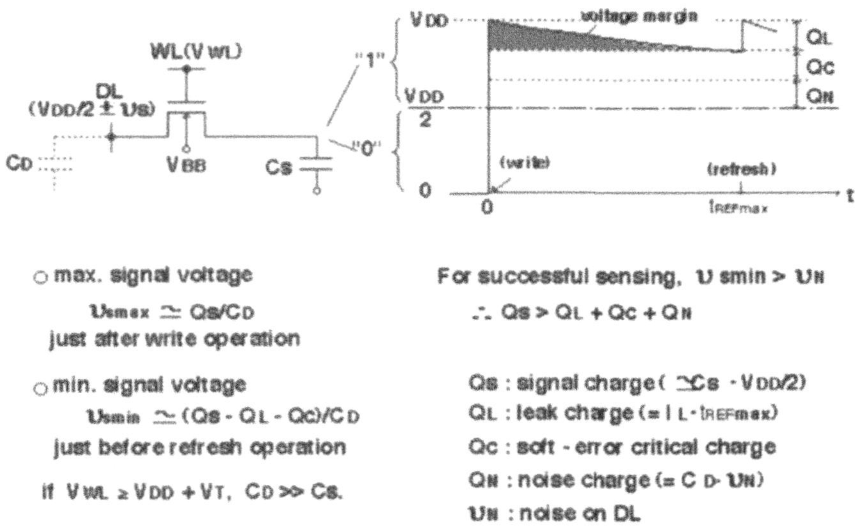

Figure 9. Requirement from read operation.

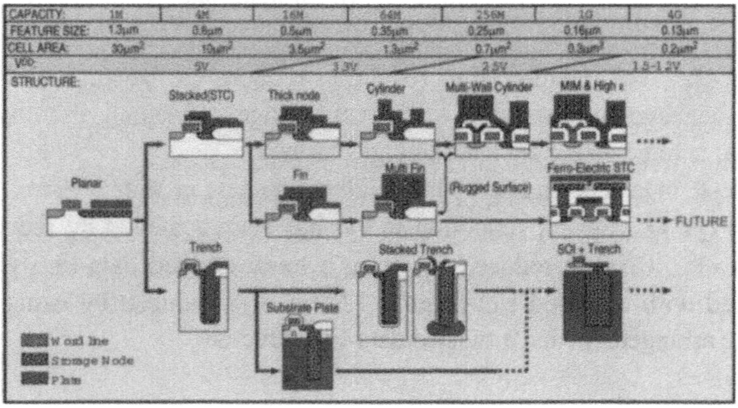

Figure 10. Trends in DRAM cell structure [7].

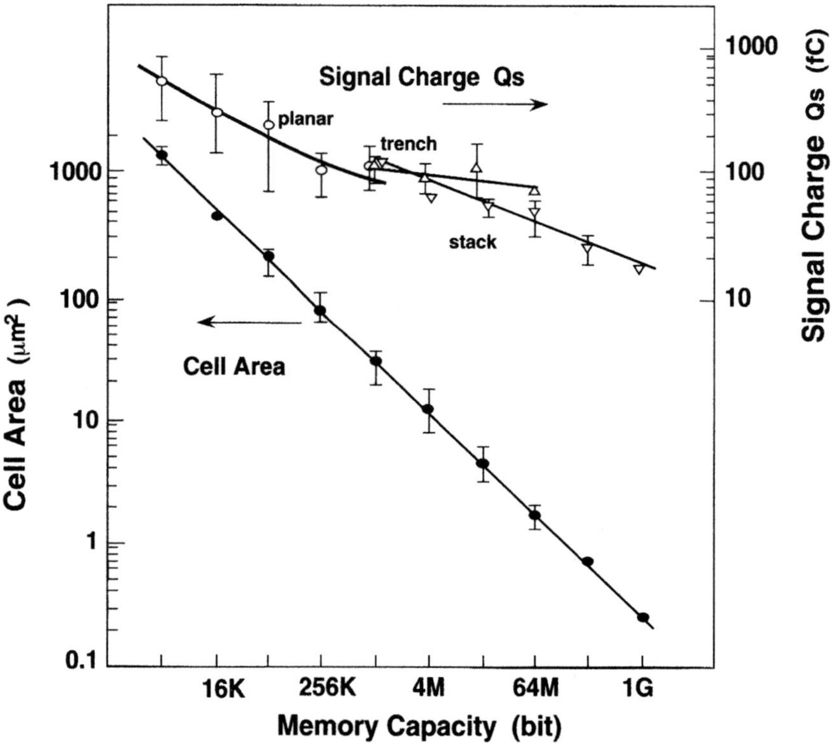

Figure 11. Trends in DRAM cell [1].

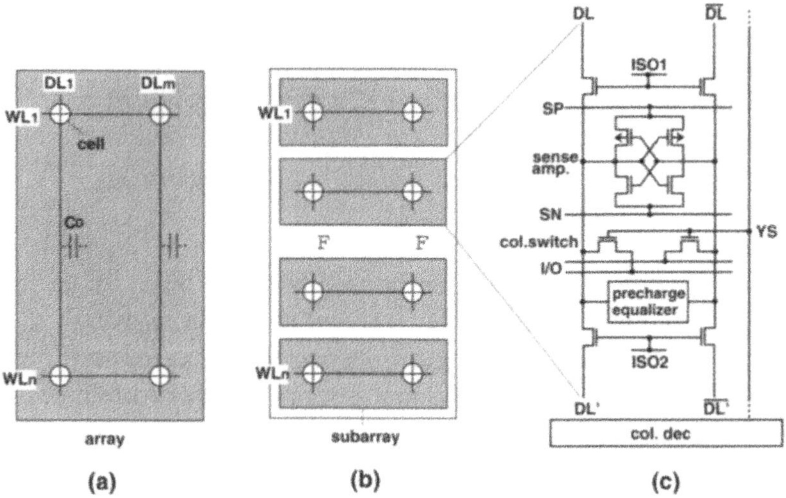

Figure 12. Concept of multi-divided data line. (a) non-divided data line. (b) multi-divided data line. (c) actual circuit.

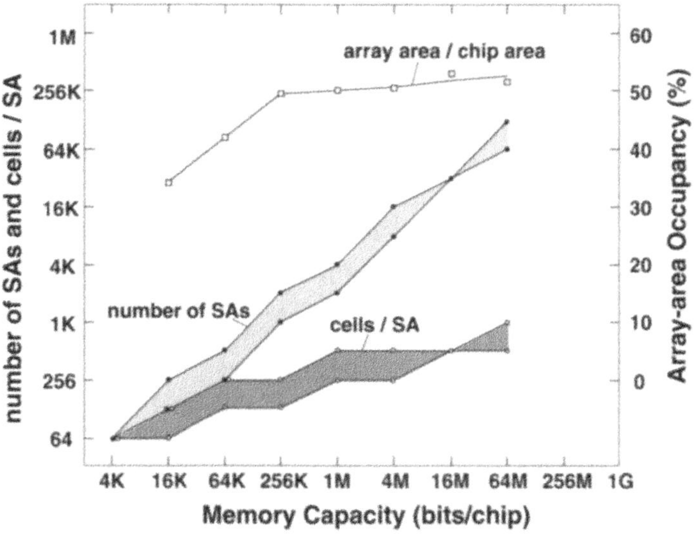

Figure 13. Trend in number of sense amplifiers and their occupancy [2][5].

Even this arrangement suffers from noise caused by capacitive imbalance between a pair of data lines. In modern DRAM design in which dummy cells have been eliminated to cut the area penalty, capacitive imbalance by Cs causes noise during sensing. Moreover the offset voltage of a sense amplifier is serious because there are numerous sense amplifiers [2,5] resulting from a multi-divided data-line structure. There are one million in a 1-Gb chip (Figure 13). Thus, a considerably large offset voltage is generated by offset-voltage deviations. The most practical solution has been the enlargement of the sense amplifier FETs at the expense of area. In any event, a higher S/N design is the key to ultralow-voltage design.

3.1.2 Write Operation

The V_T at full write operation is usually larger than that at full read operation, requiring more boosted voltage operation, although the word voltage (V_{WL}) for both full read and write operations has the same amplitude for the sake of simplicity in the actual design. For full write operation the source voltage of the cell transistor is V_{DD} for "1" write, or 0 V for "0" write. On the other hand, for full read operation it is almost a half-V_{DD}, which is the data-line precharge voltage, for "1" or "0" read. Thus V_T is largest due to substrate bias effect when "1" write occurs because the source voltage is raised to the maximum voltage (V_{DD}). The "1" write V_T and the necessary V_{WL} decrease with decreasing V_{DD}, depending on the body effect parameter (K), as shown in Figure 14. Note the existence of a fixed V_T (V_{T0}) that stems from the

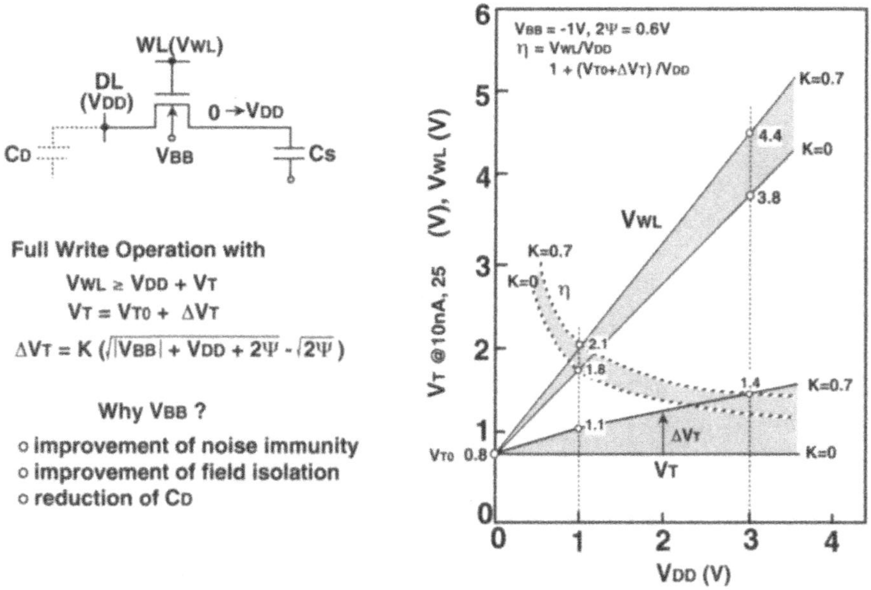

Figure 14. Requirement from write operation [5].

refresh operation as will be explained later. V_{T0} requires a relatively greater V_{WL} with V_{DD} reduction, imposing a relatively higher stress voltage on the cell transistor which calls for a higher boost-ratio (η). As an example, when K=0.7, a 2.1 V V_{WL} for a 1 V V_{DD} requires a large boost ratio of 2.1. Unfortunately, in the actual design, a greater V_{WL} is necessary because of considerably large extrinsic-V_T-variations throughout the chip. This is because the V_T of the cell transistor, the smallest transistor in the chip, is susceptible to narrow channel effects and isolation characteristics. Thus, a cell-transistor with a small K and a high η peripheral circuit have become increasingly important in ultralow-voltage design.

3.1.3 Refresh Operation

A cell that is not selected during the period of t_{REFmax} must hold data even under the worst conditions. The worst conditions are established by a combination of maximum junction temperature (Tjmax) and successive low-level disturbance from the corresponding data line. This is because Tjmax maximizes cell leakage current which is comprised of two components: the p-n junction leakage current at the cell storage node, and the transistor subthreshold current flowing from the storage node to the data line. Here Tjmax is attained by maximum ambient temperature and minimum cycle time operation. The disturbance further enhances the subthreshold current since the lowest DL voltage makes the cell-transistor VT lowest. Thus, the worst conditions are eventually successive low-level data-line disturbances due to successive operations of other cells on the data line at minimum cycle time and maximum ambient temperature. Obviously, there is a minimum VT (VT0) at a 0-V source (data line) voltage to ensure the data retention. For example, a VT0 that is defined at 10 nA and 25_ is about 0.8 V. This VT allows the cell to hold the data for a t_{REFmax} assuming an acceptable QL of 3fC and a Tjmax=100_, as shown in Figure 15. The existence of a minimum VT, which can never be scaled down, eventually makes ultralow-voltage cell-design difficult.

3.2 Subthreshold Current Reduction

The subthreshold-current reduction circuits [1,3,5] proposed so far can roughly be categorized as gate-source backbiasing schemes, utilization of multi-V_T, and dynamic V_T.

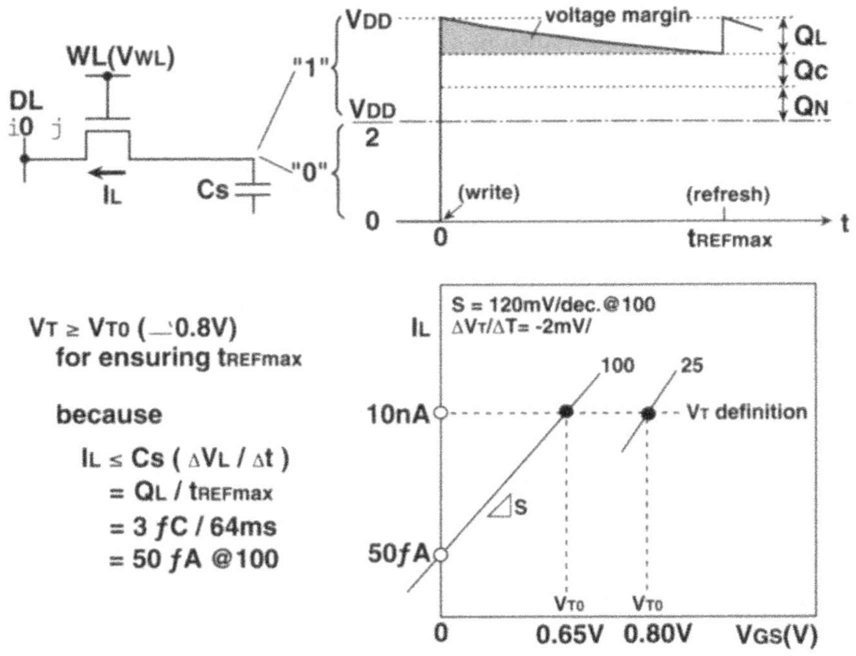

Figure 15. Requirement from refresh operation.

3.2.1 Gate-Source Backbiasing Circuit:

Two schemes for gate-source offset driving and gate-source self-backbiasing are well known. Figure 16 shows a gate-source offset driving scheme applied to a cell FET [3]. The boosted sense ground (BSG) shown in Figure 16(b) features the lowest data-line voltage raised by _VDL to create a backbias for nonselected cell FETs. The subthreshold current flow is cut off even for a low V_T as long as the sum of _VDL and V_T is larger than the minimum V_T (V_{T0}) as was previously explained. In BSG the design of a _VDL generator is as difficult as that of an on-chip voltage-down converter (VDC) as will be described later, because it must sink a large data-line discharging current. The negative word line (NWL) shown in Figure 16(c) works in a similar way. Note that the stress voltage to the FET gate insulator is relaxed if the data-line voltage is set between the lower and upper word-line voltages. Thus NWL matches lower V_{DD} operation, and reduces the gate-insulator stress voltage [8] despite difficulty in generating a stable negative word-line voltage.

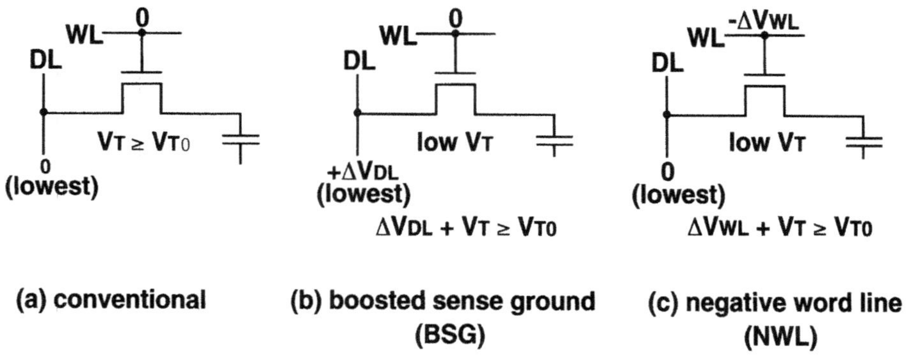

Figure 16. Gate-source backbiasing schemes applied to DRAM cell [3]. VT0: minimum VT necessary for preventing the subthreshold current flow under the data-line "L" disturbances.

Figure 17 shows another example of gate-source offset driving [3] applied to a CMOS inverter. This design enables the use of a low V_T by offsetting the source level of the driver by the difference between high and low V_Ts. Thus, it achieves high-speed switching due to reduced signal swing, while keeping the subthreshold current sufficiently low. Consequently, it is suitable for a bus driver with heavy loading capacitance although it needs two on-chip voltage converters for V_{DL} and V_{SL}.

Figure 18 shows the principle of a gate-source self-backbiasing scheme [9] using a switched-source-impedance (Sw-R) circuit. It features a switch Ss and a resistor Rs which are connected in parallel and inserted at the source of the NMOS transistor M_N, the back gate of which is connected to the ground. In order to achieve concurrent high-speed operation and low standby current, Ss is on in active mode, while it is off in standby mode. In standby mode, the subthreshold current I_L through the resistor Rs raises the source voltage V_{SL} to I_L•Rs, creating gate-source backbiasing. The current is reduced through the following two mechanisms. First, the back-gate bias of -V_{SL} enhances V_T by _V_T and the current is reduced from I_{L0} to I_{L1}. Second, the gate-source voltage of M_N becomes negative, -V_{SL}, and the current is further reduced from I_{L1} to I_{L2}. The subthreshold current is reduced by 3-4 decades using a scheme with a V_{SL} of 0.3 V. Note that negative feedback through Rs provides immunity against V_T-fluctuations which becomes larger with device down-scaling. Only one low V_T FET can realize switched impedance because Rs is regarded as the leakage resistance of Ss. This scheme is also applicable to other logic gates as long as the input voltage is predictable. Fortunately almost all RAM chip node voltages are predictable, as discussed before.

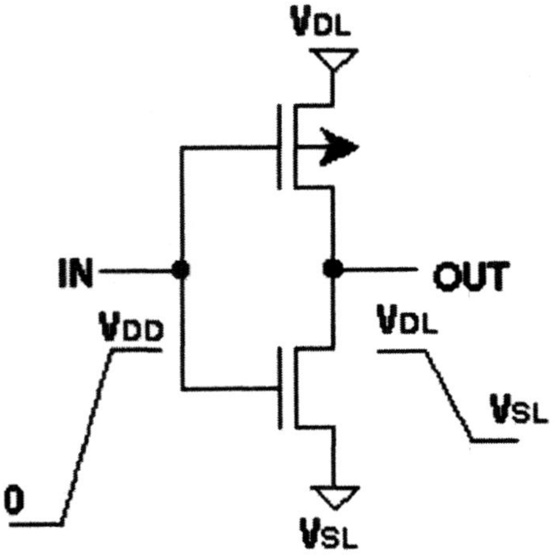

Figure 17. Offset driving inverter [3].

Figure 19 shows various applications of the scheme [9]. The scheme shown in Figure 18 is only effective when the voltage levels of terminals IN and OUT are low ('L') and high ('H'), respectively, as shown in Figure 19(a). If OUT is at a low level, switched impedance must be inserted at the source of PMOS transistor M_P as shown in Figure 19(b) because M_P is in the subthreshold region. Note that in both cases full-swing output voltages are available despite the source impedances. If OUT is at a tristate (high impedance), however, switched impedances are necessary at both sources as shown in Figure 19(c). This is because the OUT voltage level is determined by another circuit with which the output terminal is shared. Switched-impedance can be shared by other inverters to minimize the area penalty [9,10]. The most efficient application of sharing is to iterative circuit blocks [1].

Figure 20 shows switched-impedance shared with an iterative circuit block [1], that is, decoded word drivers to reduce the standby subthreshold current. A low V_T P-ch switching FET, Qs, inserted between the power-supply line V_{DD} and the driver-FETs (Q) common-source terminal, is switched impedance working as a current-limiting device. In the active period, the successive operation of selection and word-line driving is done after the power line PSL is connected to V_{DD} by turning on Qs. Here, the Qs channel width (Ws) can be reduced to an extent comparable to the Q channel width (W) without degrading speed, since only one of the n driver-transistors turns on. Just after the standby period starts when Qs is turned off, the total subthreshold current, nI, causes voltage drop (V_{SL}) at the power line, because Qs acts as impedance. As a result, the voltage drop creates a gate-source back bias to each PMOS driver-transistor so that current is reduced. The current reduction ratio, γ, and, V_{SL} are expressed in the figure. The subthreshold current decreases exponentially with V_{SL}.

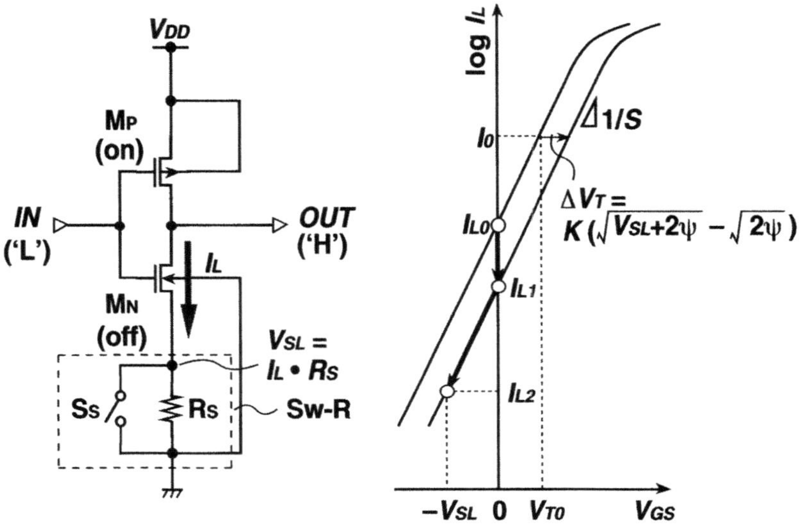

Figure 18. Principle of gate-source self-backbiasing scheme by switched source impedance circuit [9].

Obviously, the current is drastically reduced because Ws is comparable to W and the number of driver-transistors, n, is large. The descent of the PSL node stops within 200-300 mV below V_{DD} for a 256-Mb DRAM. This enables the high-speed recovery (2-3 ns) of the PSL node back to the V_{DD} level in the transition from the standby mode to the active mode. Note that if Qs has sufficiently high V_T the PSL node is discharged to 0, implying a slow recovery time, and increased charging current and power.

The subthreshold current in active mode is another concern for iterative circuit blocks, although it can be reduced in standby mode by the circuit described above. After one selected word line is activated, all the drivers are sources of subthreshold current, eventually dominating total active current. This is overcome by partial activation of the multi-divided power-line [1], as shown in Figure 21. It features a selective power-supply to part of the circuit block by being divided into m sub-blocks each consisting of n/m circuits. Operation is undertaken by turning on a switch which corresponds to a selected (activated) sub-block, while the others remain off. All the nonselected (inactivated) sub-blocks have no substantial subthreshold current since the same voltage relationship as in standby mode in Figure 20 is established in each sub-block. This reduces the current to n/m_I with an m-fold reduction. Two-dimensional selection [1] has also been reported to further reduce the current.

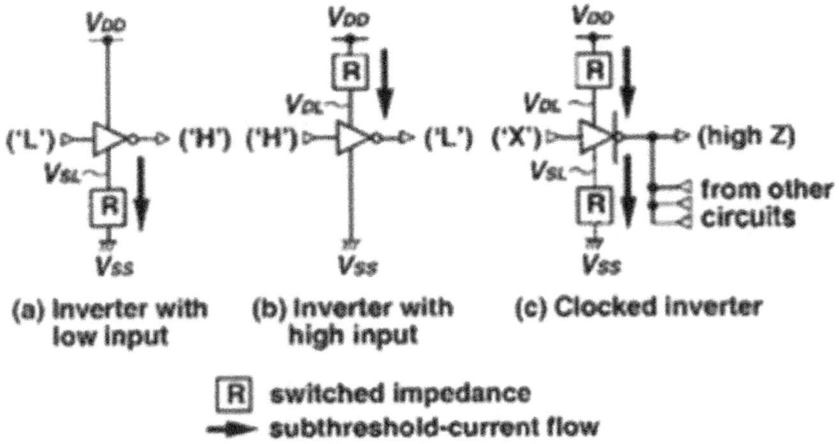

Figure 19. Variations of switched source impedance CMOS circuits [9].

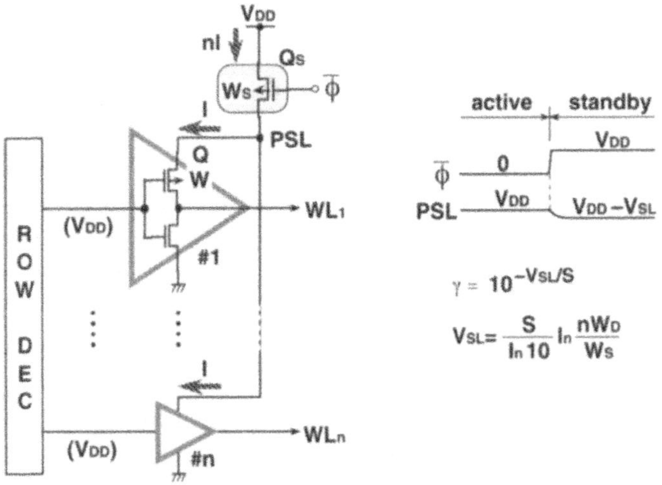

Figure 20. Application of gate-source self-backbiasing scheme to word driver [1].

3.2.2 Multi-V_T Circuit:

This cuts off the leakage path with a high-V_T MOSFET while using a low-V_T MOSFET for the main signal path during the active period. The following are typical applications for the power switch and logic circuits. Figure 22(a) shows a switched-power-supply inverter with a static level holder [1]. This is useful for some applications in which input voltage is not predictable. The power supply of the CMOS circuit is controlled by FET switches Q'_N and Q'_P. The V_Ts of all FETs except Q_N and Q_P are sufficiently high, allowing negligible subthreshold current. As soon as the input level has been evaluated at high speed as a result of the low-V_T of Q_N and Q_P and the resultant output is held in the holder, the switches are turned off. Consequently, the output level is maintained without subthreshold current. The level holder area can be minimized since it only plays a role in holding the level. The level holder and Q'_N can be eliminated if the output level does not need to be held. If this is the case, the power switch can be shared by many internal low-V_T circuits without any subthreshold current [10], as shown in Figure 22(b). However, the internal V_{DD} node and all the output nodes of the circuits are finally discharged, requiring a long recovery time and a large charging current and power. A scheme [11] to refresh the slightly degraded voltage of an internal V_{DD} line at a fixed interval solves some problems.

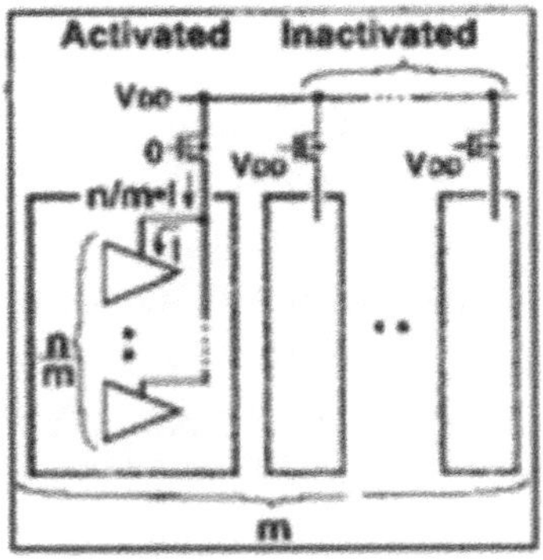

Figure 21. Partial activation of multi-divided power-line [1].

However, the interval strongly depends on V_T and its variations. Therefore power switch use is limited to the active to standby-mode control circuit which accepts considerably slow speed. Figure 22(c) shows application to a logic circuit [12]. A high-V_T prevents current leakage in the standby mode.

High-V_T eventually restricts ultralow-voltage operation although 1 V at most can be managed. This is because the conductance of high V_T FETs decreases relatively as the V_{DD} approaches high-V_T. This implies that the additional area and clock-power due to inserting FET increase relatively to compensate for conductance decrease.

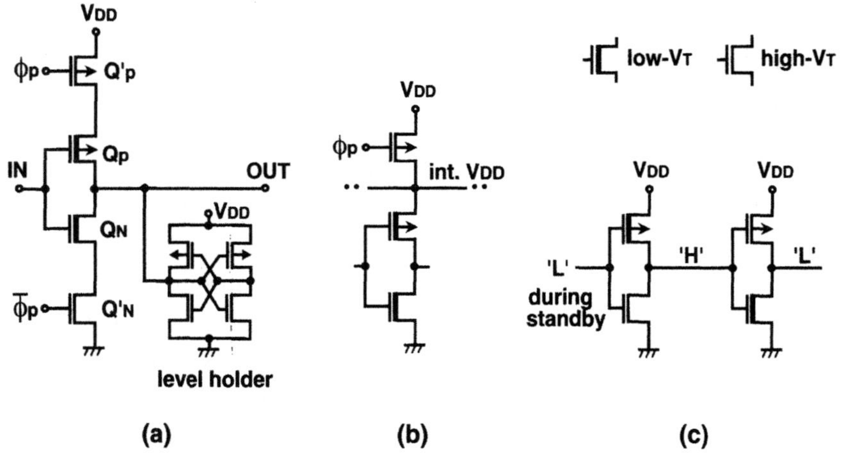

Figure 22. Multi-VT circuits [1][10][12]; (a) power switch with level holder; (b) shared power switch; (c) logic circuit.

3.2.3 Dynamic V_T Circuits

The dynamic V_T scheme is becoming increasingly important despite many possible problems which need to be solved before its application to products. This can be achieved by driving the well (or substrate) so that V_T in the inactive period is higher than it is in the active period. This offers fast operation in the active period while suppressing leakage current in the inactive period. Note that a large V_T change, which is needed for ultralow-voltage operation through well driving schemes, can be attained using a shallow substrate bias voltage (V_{BB}), a large K, and a large _V_{BB}. This is justified by the following expression:

$$V_T = V_T(0) + \Delta V_T = V_T(0)\left\{1 + \frac{K}{2\sqrt{V_{BB} + 2\Psi}} \cdot \frac{V_{BB}}{V_T(0)}\right\} \quad (4)$$

$$\Delta V_T = \frac{K}{2\sqrt{V_{BB} + 2\Psi}} \cdot \Delta V_{BB} \quad (5)$$

$$K \propto \sqrt{N} \cdot t_{ox} \quad (6)$$

where V_T (0) is the V_T for a quiet V_{BB}, Y is the Fermi potential, N is substrate doping concentration, and tox is the gate oxide thickness. Here, the challenge is to achieve a precise V_{BB} control and V_{BB} noise suppression, especially during the active period (i.e. a low V_T). Another challenge is to develop a scaled-down FET with a K as large as possible because K tends to be smaller with smaller FETs.

There have been three types of proposals based on the circuit scale: driving the common well or common source for the whole chip to simultaneously change all V_Ts in the chip, driving only the common well of a certain circuit-block in the chip, and driving the well of each individual circuit.

3.2.4 Well or Source Driving of Chip

There have been two proposals for this, namely, changing the well bias voltage (V_{BB}) while fixing a common source voltage to FETs, and changing the source voltage while fixing the V_{BB} instead. Figure 23 shows a V_{BB} driving scheme [13] for a 0.3-μm CMOS LSI. To obtain a low V_T of 0.1 V in the active mode and a high V_T of 0.5 V in the standby mode, the V_{BB} of the p-well is changed from-0.5 V to-3.3 V while the V_{BB} of the n-well changed from 1.4 V to 4.2 V to establish the same bias condition for V_{DD}=0.9 V. A large V_T change of 0.4 V is responsible for a large FET of 0.3 μm and a large V_{BB} swing, as explained previously. V_{BB} in the active period is well regulated by low frequency pumping controlled by leakage current monitors (LCMs) after completing high-speed well discharge by high frequency pumping. In the sleep (standby) mode high frequency pumping again discharges the well to a deep V_{BB}. At the beginning of the transition from sleep to active mode a substrate injector consisting of a CMOS inverter quickly charges up the well to-0.5 V. In principle, VBB driving can inherit traditional circuit and design-methodology because only VBB control is required. In addition to the IDDQ test, adjustment the chip-to-chip VT variations by an appropriate VBB setting can be achieved.

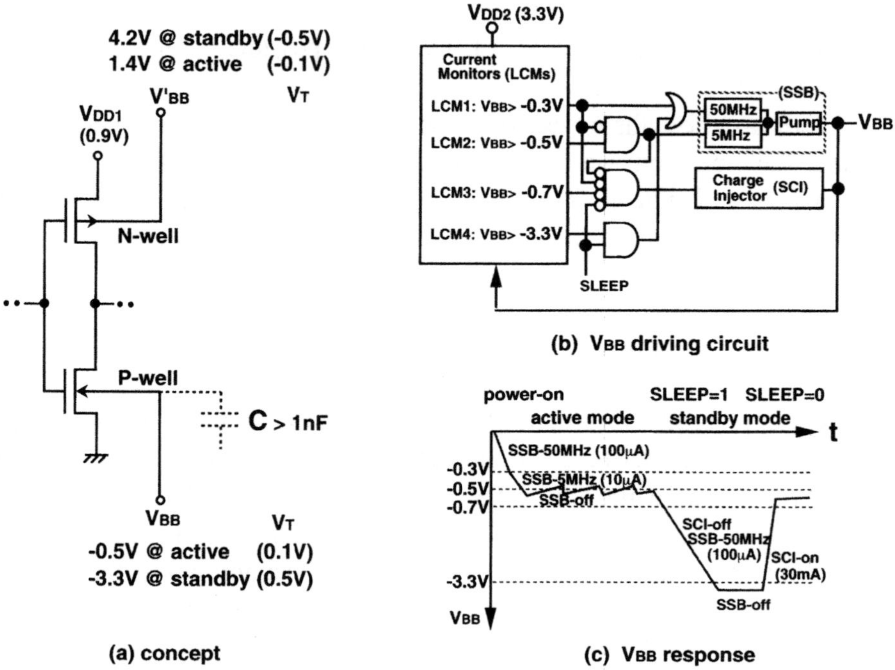

Figure 23. Dynamic VT scheme by well driving [13].

The VBB response time, however, is as low as a few hundreds of ms. This is because on-chip VBB generation (Figure 27) cannot quickly drive well capacitance heavier than 1nF with a VBB swing as large as 2.8 V. This slow response prevents the reduction of subthreshold current in the active period. A large VBB swing also needs an additional external supply voltage of 3.3 V, which excludes it from a single voltage supply scheme. The ac and dc instabilities of VBB that originate from a floating substrate (well) may be caused by the voltage bump of two external power-supplies, by coupling noise through junction capacitance, and by the increase in the substrate (well) current (IBB) of NMOSFETs, especially in high speed designs using short channel FETs. Possible source-well forward bias caused by instability makes chip operation hazardous. Note that in modern DRAM, as in MPUs and ASICs, the tight connection between the source (that is, a power line of VDD or 0 V) and the well in each MOSFET never creates a source to well voltage-difference, ensuring a fixed VBB of 0 V throughout in the chip. Possible power-on rush current or CMOS latch-up [5] at an extremely low VT, developed by intermediate VBB during power-on, is another concern.

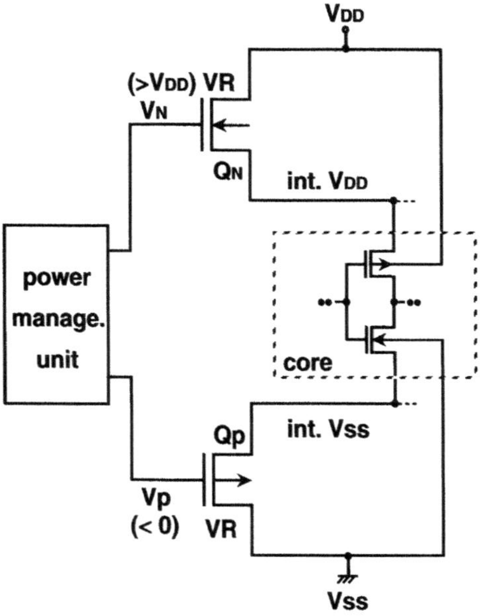

Figure 24. Dynamic VT by source driving [14].

Figure 24 shows a source-driving scheme [14] under a fixed well-bias voltage. In the standby (sleep) mode, the common source voltage of NMOSFETs in the internal circuit is raised while that of PMOSFETs is lowered to increase V_T. This is accomplished by controlling the gate voltage of the output MOSFETs (Q_N, Q_P) of the voltage regulators (VR) which are inserted into the main path of the power supply (V_{DD} to V_{SS}). The design-parameter, temperature and voltage variations are automatically compensated by precisely controlling the gate voltages, which is accomplished by the use of a CMOS delay line, a phase detector and charge pumps in the power management unit. However a pulsive source noise, developed when operating the internal core circuit, would be a major obstacle. This is because Q_N and Q_P cannot manage a large current without area penalty. The slow response time of the internal power lines when switching mode, such as V_{BB} driving, is another problem. In addition, the V_{DD} must be higher than the internal supply voltage, thus limiting low V_{DD} operation.

3.2.5 Well Driving of Circuit Block

Figure 25 shows a common well driving scheme [15] applied to a sense-amplifier block that suffers from inherently slow speed, as explained previously. This scheme combined with a triple-well structure achieves a low VT without well-bias during amplification so that speed is increased. A high VT is attained by applying enough bias just after sensing/restoring operation. Here NMOSFETs are located in a P-well isolated from the P-type substrate by an N-well and N-bottom while PMOSFETs are located in the N-well. In the equalizing period the wells return to the half-VDD level. The resultant low VT boots the current so that the data-lines are quickly equalized.

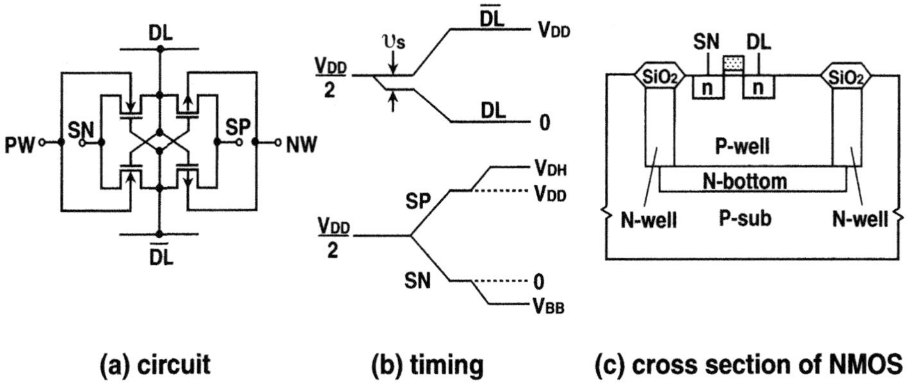

Figure 25. Sense-amplifier well driving [15].

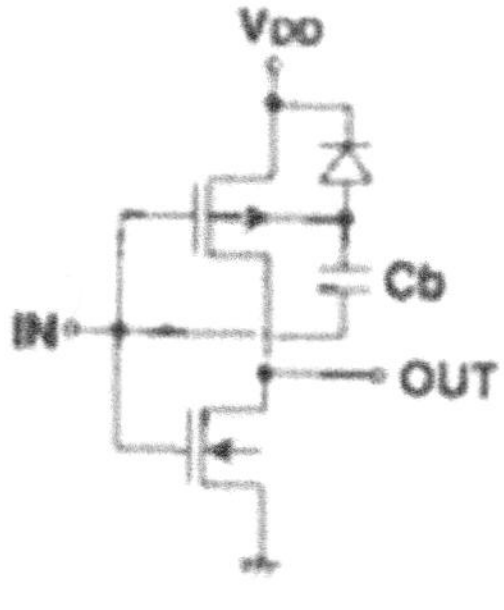

Figure 26. Capacitor-coupled well driving [16].

3.2.6 Well Driving of Individual Circuit

Figure 26 shows a well driving [16] for an individual circuit. It features capacitor-coupled driving in which the MOSFET well is dynamically connected to the gate by a capacitor so that VT is automatically adjusted during operation. Hence it enables a much higher current when turned on (effective low VT) and much lower leakage current when turned off (effective high VT). Diode is used to discharge Cb. Automatic VT control is important in designing a complicated chip, since all the design issues involving the subthreshold current are confined to the individual circuit level. However, there are drawbacks in terms of the increase in loading capacitance for the previous stage, and area penalty due to an additional capacitor.

3.2.7 Suppression of or Compensation for Design Parameter Variations

In addition to stringent control of channel length, a shallow junction MOSFET, which is formed by reducing of ion-implantation energy and process temperature, reduces the V_T variations and offset voltage of sense amplifiers[3]. Compensation circuits against design-parameter variations are also important. On-chip voltage generators which track the variations [6] may be one solution. The voltage-down converter which will be discussed later can regulate internal supply voltage against extreme external V_{DD} variations. In addition, internal supply voltage could be adjusted based on variations in design parameters through adjusting the on-chip reference voltage. The substrate (or well) -bias generator could control V_T. The voltage-up converter could generate raised output by tracking the cell V_T.

3.2.8 Single Power-Supply and Power-Supply Standardization

In addition to a half-V_{DD} generator, a substrate bias (V_{BB}) generator, a voltage-up converter (VUC), and a voltage-down converter (VDC) are important single power-supply circuits. The VDC is also essential in standardizing power supply. All generators and converters require high conversion-efficiency, precise control and trimrning of internal voltages based on design-parameter variations. Their power consumption is another concern. For example, a higher boost-ratio, as explained earlier, for a voltage-up converter consumes higher power because of its lower conversion-efficiency. Thus low-power/low-voltage analog circuits are expected to become increasingly important.

A V_{BB} generator is indispensable for the stable operation of a DRAM, especially for the array. This provides a negative DC voltage to the P-substrate which is almost capacitive. Historically, both a high V_T of over 0.5 V and a deep V_{BB} of -2 to -3 V for NMOSFETs have ensured stable chip operation with a negligibly small V_T change (_V_T) despite a large V_{BB} bounce (_V_{BB}) of about 1 V and a large K that originates from large FETs, as shown by eqs. (4) to (6). Figure 27 shows a V_{BB} generator featuring two sets of charge pump circuits: a slow cycle ring-oscillator 1 to supply a small current during retention and stand-by modes and a fast cycle ring-oscillator 2 to supply a sufficiently large current during the active cycle or when the level monitor detects that the V_{BB} level is high.

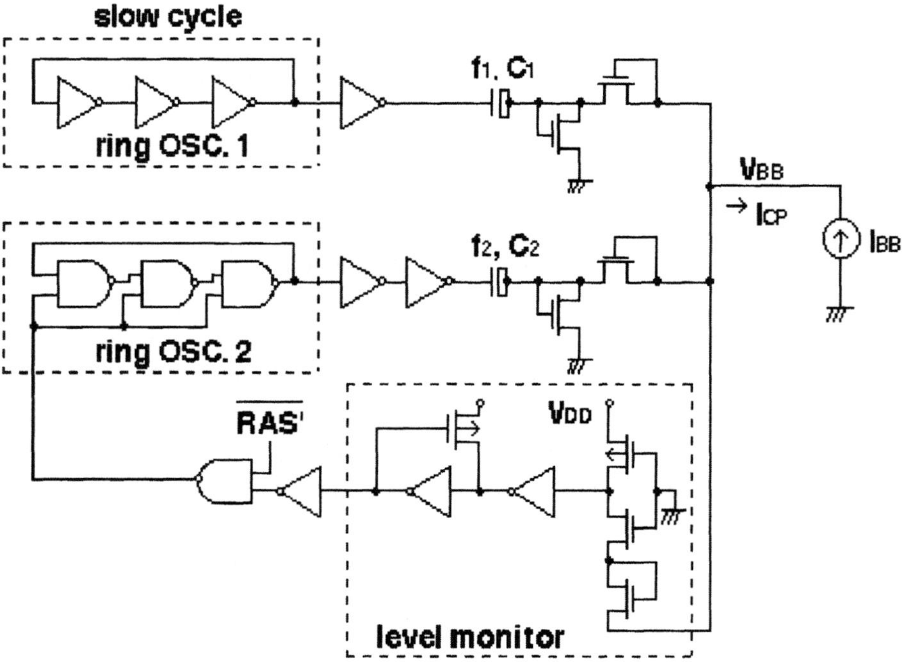

Figure 27. Substrate bias voltage (VBB) generator [4][5].

Thus, it minimizes the retention current by shutting down the fast cycle circuit. A similar approach is useful for VUC and VDC designs, as will be explained later. Another alternative is to stop the oscillation of the V_{BB} generator while the DRAM is not in an active cycle. This configuration is also useful for VUC and VDC designs. The V_{BB} generator does not compensate for pulsive noise, but compensates for quasi-static V_{BB} change due to the generator's poor charge-injection capabilities.

The on-chip voltage-up converter (VUC) [1] shown in Figure 28 provides a raised power-supply voltage (V_{DH}) to eliminate V_T drop. It has two kinds of charge-pump circuits to create charges for a pure capacitive output load: a main pump and an active kicker. The main pump compensates for small charge loss caused by leakage current of the load. It is driven by a ring oscillator which is activated when V_{DH} is lower than the level determined by the level monitor. The active kicker operates synchronously with load-circuit operation such as ISO driving (Figure 12), word-line driving, and output buffer driving. This circuit compensates for large charge loss caused by load circuit operations. Non- operation of the main pump and extremely slow cycling of the active kicker in the data-retention mode create a minimized retention current.

An on-chip voltage-down converter (VDC) [1] offers single-and standard-V_{DD} operation by tailoring the internal supply voltage based on the breakdown voltage of the MOSFETs in the internal core circuit.

It also provides a low power with higher speed and smaller chip area [4]. The keys to designing VDC are provision of a stable and accurate output voltage under rapidly changing load current and provision of on-chip burn-in capability. Figure 29 shows a schematic of a typical VDC and the step response for the load current, I_L. The almost fixed output voltage, V_{DL}, is about 3.3 V for V_{DD}=5 V, 16-Mb DRAMs. For accuracy and load current driving capability, it has a current-mirror differential amplifier (Q_1-Q_4,Q_s) and common-source drive transistor (Q_6). The array current for a DRAM is fairly large compared with that of an SRAM. The peak height is more than 100 mA with a peak width of around 20 ns.

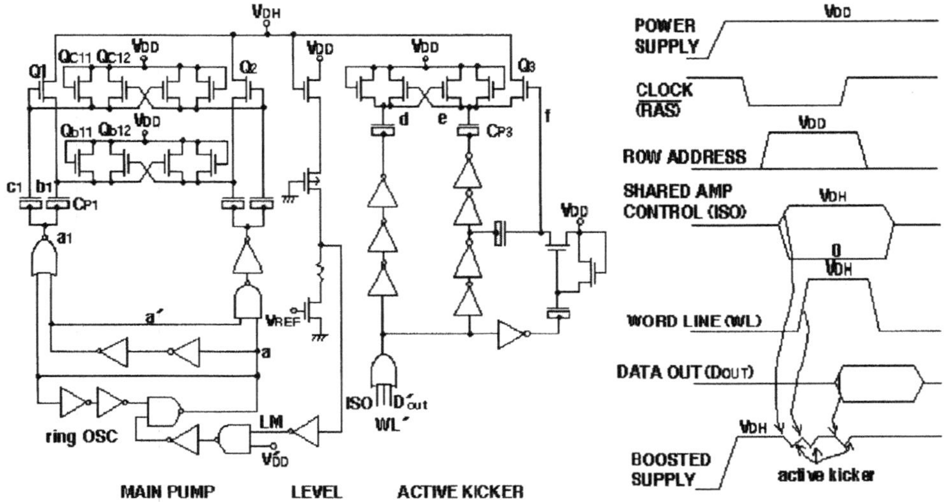

Figure 28. On-chip voltage-up converter [4][5].

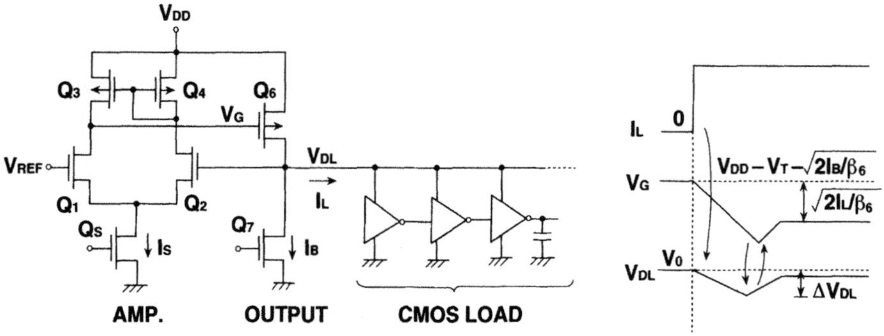

Figure 29. On-chip voltage-down converter [1].

Thus, the gate width of Q_6 has to be more than 1000 μm. In order to minimize the output voltage drop _V_{DL}, the gate voltage of Q_6, V_G, has to respond quickly when the output lowers.An amplifier current Is of 2 to 3 mA enables such a fast response time. Bias current source I_B is needed to clamp the output voltage when the load current almost reaches zero. To ensure sufficient loop stability with minimized area and operating current, phase compensation is indispensable. The reference voltage V_{REF} must be accurate over wide variations in V_{DD}, process, and temperature for stable operation, because the voltage level determines the amount of cell signal charge as well as speed. A band-gap V_{REF} generator and a CMOS V_{REF} generator utilizing threshold voltage difference have been proposed to meet requirements. Burn-in operation with the application of a high stress voltage to devices is indispensable in VLSI production both in terms of reliability testing and chip screening. To achieve this purpose, the VREF generator is designed to output a raised voltage when the V_{DD} is higher than the value required for normal operation. Otherwise, the fixed voltage fails to apply a higher stress voltage. The VDC increases the area and current of a 16-Mb chip by less than 1% and about 3%, respectively.

Using these generators and converters a gradual transition toward external or internal power-supplies of sub-V levels seems inevitable in terms of ever-decreasing devices, as shown in Figure 30. Thus a landmark will be passed at around 1 V which is suitable for one cell battery-operation. At the 0.5-V level, even one solar-cell operation may be possible. Customized RAMs in which the first priority is on ultralow-power rather than standardization, however, will not use an on-chip converter to avoid power loss in the converter.

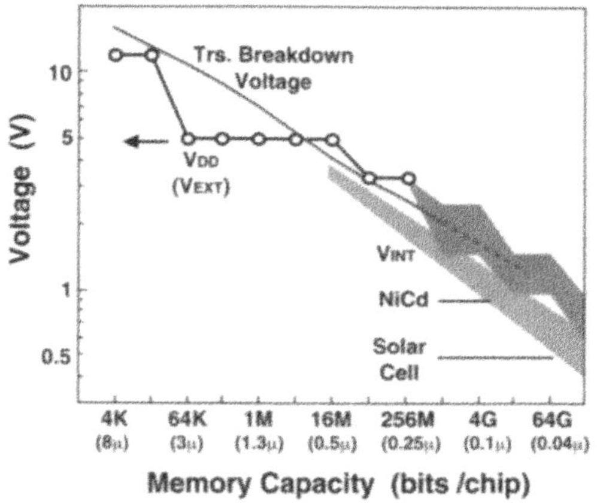

Figure 30. Standard power supply voltage (VDD) of DRAMs [3].

4. ULTRALOW-VOLTAGE SRAM CIRCUITS

In ultralow-voltage operation, the SRAM array is a major concern in terms of subthreshold current and V_T mismatch. It is the largest channel-width block dominating the subthreshold current of the chip, unlike the situation in DRAMs. In addition, it has the largest number of flip-flop circuits whose operations are sensitive to V_T mismatch. Note that the current in the remaining circuit blocks could be suppressed with circuits similar to those of DRAMs previously described. A detailed discussion is in Ref. [6].

A high-V_T is needed for cell driver-FETs to suppress huge subthreshold-current, as discussed before. A high-V_T is also necessary for cell transfer FETs to avoid leakage current flowing to either of the data lines. A high V_T, however, decreases the cell voltage-margin as the cell-supply voltage (V_{DD}) approaches the V_T, limiting the minimum V_{DD} of the chip. Figure 31 shows various cell-driving schemes to overcome this limitation. Two-step word driving [17] enables high speed write-operation directly from the data lines by boosting the word line to eliminate the V_T-drop of transfer FETs.

The poor driving capability of TFT cell loads needs the above word driving despite a write-speed penalty caused by the necessity of a preceding read-operation. The raised dc voltage [18], generated by an on-chip voltage-up converter to supply cell loads, allows the storage node voltage of the cell to quickly rise during write-operation because of increased TFT conductance. Both schemes, however, suffer from slow read-operation because of the high VT of transfer FETs.

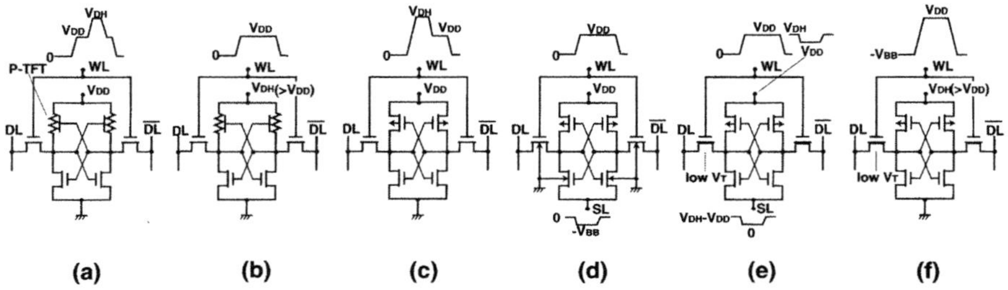

Figure 31. Various SRAM cell-driving schemes. (a) Two-step word-voltage [17]. (b) Raised VDD (=VDH) [18]. (c) Step-down boosted wordline [19]. (d) Negative source-line [20]. (e) Offset source-line [21]. (f) Boosted storage node [22].

The step-down boosted word line [19] offers high-speed operation as well as low power. However, additional delay in boosting, and variations in the amplitude and duration of the boosted pulse caused by process variations are involved. The negative source-line scheme [20] also achieves high speed with a reduced VT in cell driver-FETs and a boost effect for the cell transfer-FETs. However, heavy capacitance, which almost equals data-line capacitance multiplied by the number of selected cells, established at the source-line prevents single VDD operation. This is because an on-chip negative-voltage generator comprised of charge-pumping circuits can never cope with such heavy capacitance. The offset source-line scheme [21] solves the heavy capacitance issue, requiring an on-chip voltage-up converter instead of the above negative-voltage generator. In the above cells the minimum VDD may be around 1 V in practical design in which soft error, VT variations and VT mismatch in addition to high-VT are considered. In particular, VT mismatch that continues to increase between paired FETs in a cell along with FET miniaturization may limit low voltage developments in the future. Hence the boosted storage-node scheme [22] is effective in lowering the minimum VDD down to less than 0.5 V with the help of an on-chip voltage-up converter. A low-VT for transfer FETs combined with negative word-line biasing also helps high-speed operation. However, the challenge is to achieve sufficiently high voltage generation at low power. Obviously, voltage generation is not needed if a higher externally supplied voltage becomes available.

5. PERSPECTIVES

In this section two emerging issues are reviewed. They are an SOI structure which works even at an extremely low voltage of around 0.5 V, and intrinsic fluctuations in design parameters that continue to increase towards 0.1 μm or less, which will eventually limit ultralow-voltage operation.

5.1 SOI CMOS Technology

SOI CMOS technology may meet the requirements of ultralow-voltage operation and / or Gigabit DRAMs better than the bulk CMOS technology [23] discussed thus far. The small junction areas of the source and drain that are completely isolated by the SiO_2 layer, as shown in Figure 32, reduce junction capacitance, leakage current, and the critical charge of soft error. The small body (substrate) of each FET that is formed between only the thick SiO_2 layer and the gate area, results in small subthreshold swing (small S-factor), less short-channel effects, and a reduction in the back-gate-bias effect [25]. Less capacitance in the SOI body favors dynamic V_T control, which is attained by body control, and this lowers the operating voltage. In addition, the resulting FET structure is simple and free of latch up. Nevertheless, some problems still remain unsolved. These are: The floating body effect of a DRAM-cell FET which degrades data retention and soft error characteristics; the effect of the FET structure which inherently increases thermal resistance; and the formation of a damage-free structure and a low-cost preparation of a thick SiO_2 layer. The following is a summary of state-of-the-art SOI circuit technology.

Figure 33 shows a 3.3-V 0.6-μm floating body 64-Kb DRAM test chip which was fabricated on a SIMOX (Separation by Implanted OXygen) wafer. Oxygen was implanted at an energy of 190 KeV and at a dose of $1.8\_10^{18}/cm^2$, and high-temperature annealing was performed at 1320_ for 6 hours. The reduced C_D/Cs, which arose from reducing junction capacitance in the data line, increases the read signal voltage by 25%. Access time was also improved by about 35% at 3 V due to the reduced junction capacitance and back-gate-bias effect in the peripheral circuits.

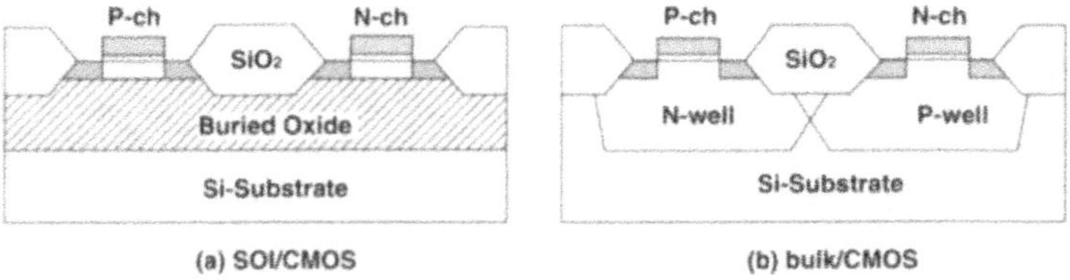

Figure 32. SOI CMOS vs. bulk CMOS [23][24].

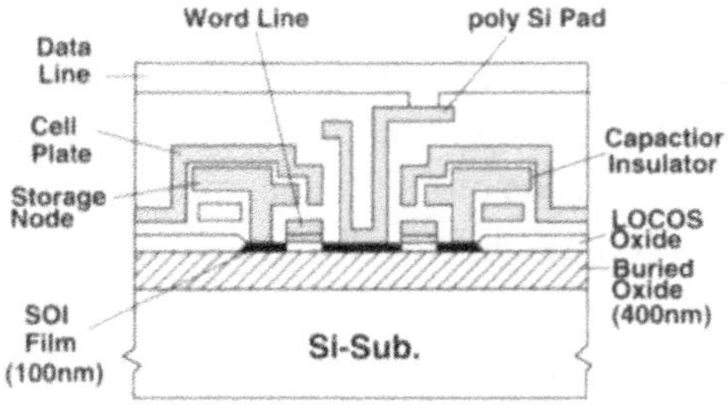

		SOI	BULK
C_D/C_S (128cells/DL)		9 (216/24fF)	12 (288/24fF)
cell signal	3.0V	150 mV	115 mV
	1.5V	58 mV	38 mV
access time	4.0V	52 ns	70 ns
	3.0V	66 ns	98 ns
	2.3V	85 ns	—
active current (3V,260ns cycle)		1.1mA	3.2mA

Figure 33. A stacked-capacitor SOI DRAM cell and experimental results of 64Kb chip compared with a bulk CMOS counterpart [23].
memory cell: 5.12μm^2, VTn = 0.8V, Ln = 0.7μm
peripheral: VTn/VTp = 0.43V/-0.6V, Ln/Lp = 0.6/0.6μm.

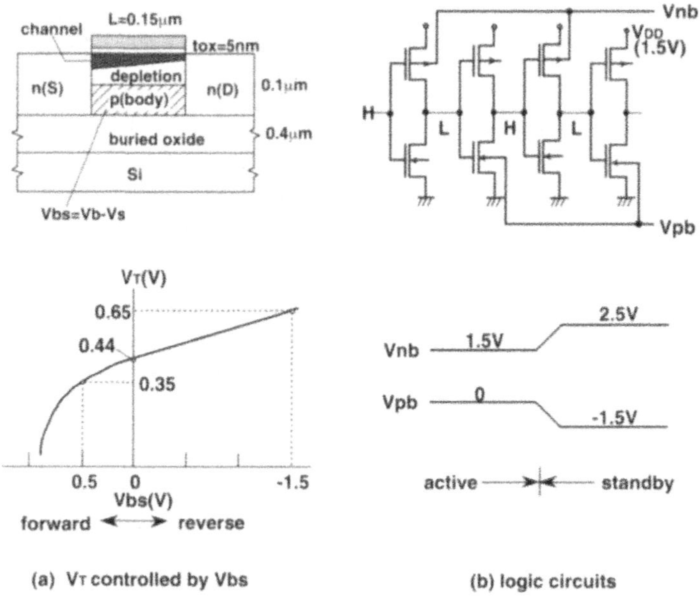

Figure 34. Body-bias control technique [24].

It has also been reported that at 2 V supply voltage an SOI CMOS SRAM operates twice as fast as its bulk CMOS counterpart [26]. A body-bias control technique to achieve a dynamic V_T further lowers the minimum operating voltage down to 1.5 V or less [24,27]. Figure 34(a) shows an NMOSFET for a hypothetical 1.5-V 4-Gb DRAM. The FET is partially depleted, and thus V_T can be controlled by body-bias voltage. Note that even when the voltage difference between the body and the source, Vbs, is +0.5 V, and the p-n junction is forward biased, current flow is negligible. Figure 34(b) shows the body-bias control logic applied to peripheral circuits. During a standby cycle, the p-body Vpb is set to -1.5 V and V_T is set to 0.65 V.

The dynamic V_T scheme can also be applied to the sense amplifier, as in the bulk CMOS previously described. The expected current of the peripheral circuits was approximately 1/20th that of the bulk, and the access time of the SOI was 35% faster than that of the conventional bulk. Such high performance has been verified using an actual 0.5-μm CMOS/SIMOX 16-Mb DRAM [27]. Body-bias control attained a high-speed access time of 46 ns even with an operating voltage as low as 1 V. The body-current clamper, which suppresses body current due to body-bias overshooting, and a negative voltage word-line scheme contribute to stable memory-chip operation.

The high speed control of body-bias in the logic is difficult since their bodies are connected, and thus the resulting body capacitance and resistance are consequently heavy despite the SOI structure. A dynamic-V_T MOSFET (DTMOS) built on an SOI [28] may solve this problem because of the control needed in individual circuits as shown in Figure 26. In it, the body is connected to the gate, as shown in Figure 35, and it has a low V_T when the FET is turned on for high current. The DTMOS can even achieve 0.5-V operation [29] in the feedback buffer in the figure, but its operating voltage is strictly limited to less than around 0.8 V due to the forward bias of the body-source p-n junction. This drawback is overcome by the power-switch FET (Qp) [30] shown in Figure 36(a). As a result of forward bias at the pn-diode, a large leakage current flows from the body to the gate if the V_{DD} is over 0.8 V, which is the diode built-in potential, as shown in Figure 36(b). Inserting a reverse-biased low-V_T MOS-diode between the body and the gate permits a higher V_{DD}, as shown in Figure 36(c). The low-V_T diode is one-tenth the Qp size. For V_{DD}>0.8 V, the diode clamps the forward bias of the Qp pn-diode, suppressing gate-leakage current. Thus, in the active period at low level SL Qp provides V_{DD} to the internal V_{DD} line with small leakage current, while in the standby period at high level SL Qp completely isolates the internal V_{DD} line from the V_{DD} line without any subthreshold current.

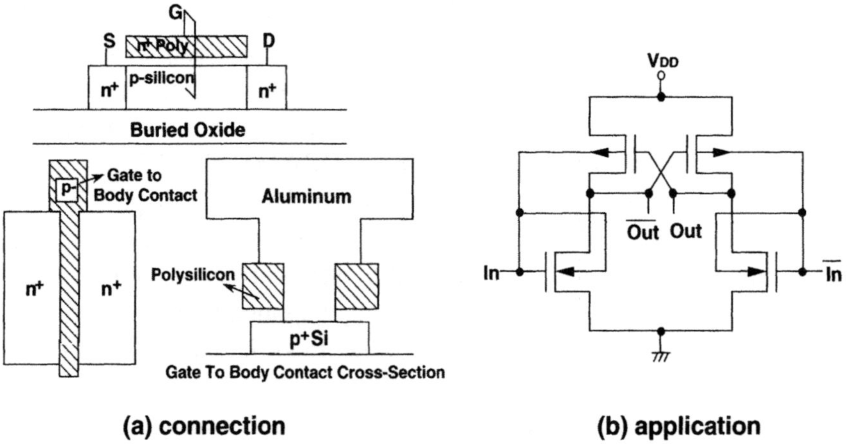

Figure 35. DTMOS concept featuring body tied to the gate [28] and its application to a 0.5V PMOS feedback buffer [29].

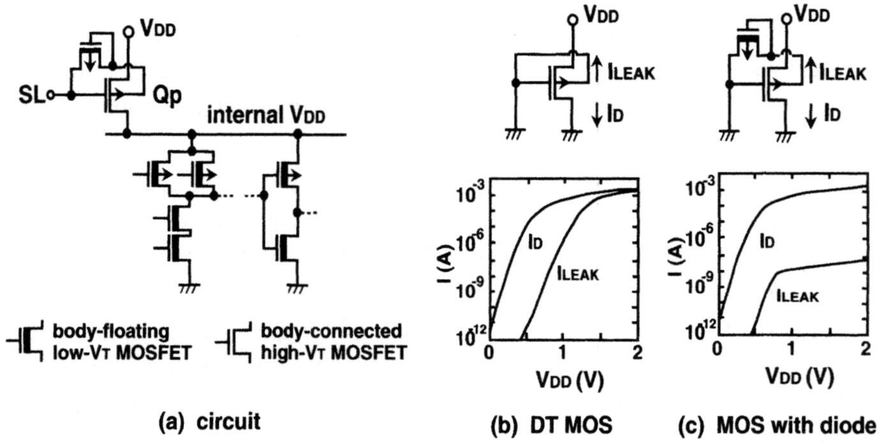

Figure 36. Multi-VT circuit applied to power-switch [30].

The major problem in the SOI structure is closely related to the floating body effect in the DRAM cell transistor: The instability of the floating body potential degrades the data-retention characteristics and soft-error immunity. Figure 37 shows the degradation mechanism for data-retention characteristics [23,31]. The data-retention time is defined as the time until the cell voltage stored by the write operation decays to almost half V_{DD}. The decay stems from the stored charges (holes) lost by p-n junction leakage (1) at the cell stored node. The resulting accumulated holes at the body raise the retention mode is related only to p-n junction leakage, while the dynamic-retention mode corresponding to the "L" data-line disturbance explained previously is related to both the p-n junction leakage and the subthreshold leakage. In the static mode, an SOI DRAM cell achieves a superior data-retention time of 550 sec at 25_, which is 6 times longer than that of a bulk memory cell [32]. In this mode the cell transistor is completely cut off despite the reduced V_T, because the data line remains at a high level of a half V_{DD} (precharge voltage of data line) and the word line is maintained at a low level of 0 V during retention. Unfortunately, the dynamic mode shortens the data-retention time for the static mode (530 sec) to 42 sec, as shown in Figure 37. The boosted sense-ground (BSG) scheme described previously improves the characteristics to some extent, raising the lowest data-line voltage from 0 V to 0.5 V. The body-refresh scheme [33] combined with the BSG can produce a long date-retention time, although additional refresh operations for the body node are necessary.

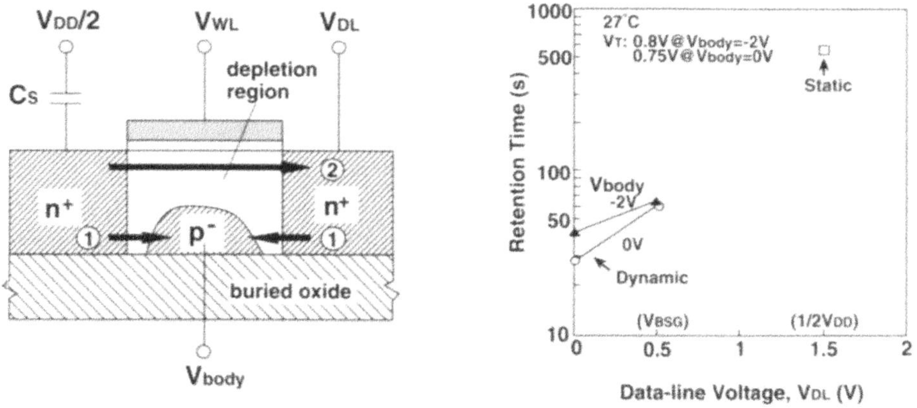

Figure 37. Data-retention characteristics of SOI DRAM cell [31]. A VDL of 1.5V (=1/2VDD) is for the static mode, while VDL of 0V and 0.5V are for dynamic mode and BSG scheme, respectively.

The floating body also degrades the soft-error immunity of MOSEFTs not only in DRAM and SRAM cells [23,34], but also in peripheral circuits. Figure 38 shows variations in simulated body potentials and leakage current over time after an _ particle incidence for floating and body-fixed SOI MOSFETs. Electrons which are generated in the floating body diffuse to the source and drain, while holes remain in the floating body region and raise the potential. This potential increase causes a large continuous subthreshold current.

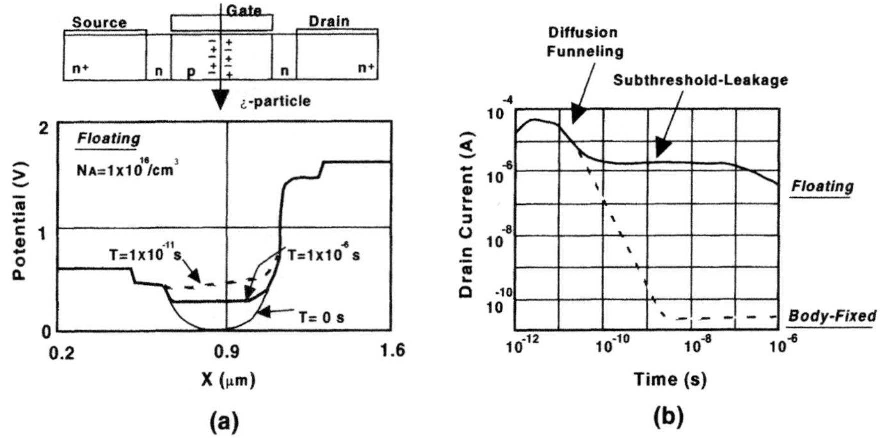

Figure 38. Potential increase (a) and resulting subthreshold leakage current (b) caused by an a particle incidence to the body region of an SOI transistor [23]. L = 0.5mm, W = 2mm, tSOI = 100nm and $tSOI_2$ = 400nm.

The introduction of body contacts [23] to suppress body-potential change is very effective in peripheral circuits. For memory cells, however, it increases the memory cell area, especially in DRAMs._

B. Random MOSFET Parameter Fluctuations

Suppressing MOSFET parameter fluctuations is essential to achieve ultralow-voltage operation, as previously discussed. Unfortunately, however, even in the absence of extrinsic dimensional variations intrinsic fluctuations in V_T, the S-factor and drain current rapidly increase as device dimensions are scaled down. The cause is random microscopic fluctuations in the number and location of dopant atoms in the channel region of a MOSFET [35,36]. Figure 39 shows the standard and maximum deviations for V_T throughout the generation. Small channel width (W=L) is for

memory cell while wider one (W=20 L) is for peripheral circuits. Approximately ±45 (W=20 L)–89(W=L)% maximum deviations from the target V_T of 0.30 V are predicted for 0.10μm chip generation. They represent unacceptably large tolerances in this critical parameter for an ultralow-voltage of 1.2 V, thus new device designs which will minimize these fluctuations are required.

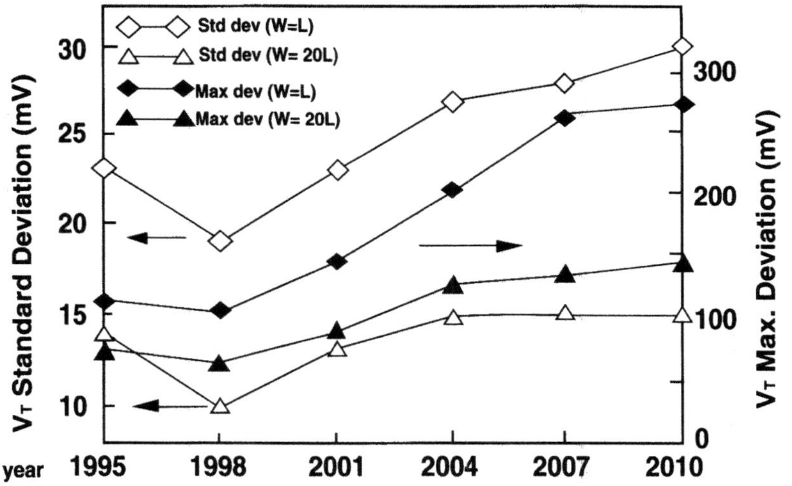

year	1995	1998	2001	2004	2007	2010
L(μm)	0.35	0.25	0.18	0.13	0.1	0.07
tox(nm)	10	5	4.5	4	3.5	3
VDD(V)	3.3	2.5	1.8	1.5	1.2	0.9
Trs. count (logic-memory)	12-64M	31-256M	93M-1G	225M-4G	450M-16G	1.3-64G

Figure 39. Standard and maximum VT deviations of MOSFET across technology generations [35][36].

6. CONCLUSION

Ultralow-voltage RAM circuits were reviewed. Through the discussion it was clarified that higher S/N cell designs, subthreshold-current reduction circuits for both active and standby modes, suppression or compensation circuits for expected wide design-parameter variations, and single-power and its standardization were keys in achieving ultralow-voltage. A state-of-the-art SOI circuit, and the emerging issue of unacceptably large design-parameter fluctuations expected in the 0.1-μm LSI era were also discussed.

7. ACKNOWLEDGMENT

The author wishes to thank K. Ishii for her helpful discussion.

8. REFERENCES

[1] K. Itoh et al., "Trends in low-power RAM circuit technologies," IEEE Proc., vol. 83, pp. 524-543, Apr. 1995.

[2] K. Itoh, "Trends in megabit DRAM circuit design," IEEE J. Solid-State Circuits, vol. 25, pp. 778-789, June 1990.

[3] K. Itoh et al., "Limitations and challenges of multigigabit DRAM chip design," IEEE J. Solid-State Circuits, vol. 32, pp. 624-634, May 1997.

[4] K. Itoh , "Low power memory design," in *Low Power Design Methodologies*, J. M. Rabaey and M. Pedram, Eds. Norwell, MA: Kluwer, Oct. 1995, pp. 201-251.

[5] K. Itoh, *VLSI Memory Design* (in Japanese), Baifukan, Nov. 1994.

[6] *Low Power CMOS Design*, R. Brodersen and A. Chandrakasan, Eds.: IEEE Press Book, 1997.

[7] A. Koike, "Key issues in manufacturing of Giga era," in 1996 VLSI Technology Workshop Digest.

[8] T. Tsuruda et al., "High-speed/high-bandwidth design methodo-logies for on-chip DRAM core multimedia system LSI's," IEEE J. Solid-State Circuits, vol. 32, pp. 447-482, March 1997.

[9] M. Horiguchi et al., "Switched-source-impedance CMOS circuit for low standby current giga-scale LSI's," in Symp. VLSI Circuits Dig. Tech. Papers, May 1993, pp. 47-48.

[10] S. Mutoh et al., "1V high-speed digital circuit technology with 0.5mm multi-threshold CMOS," in IEEE ASIC Conf. Dig. Tech. Papers, Sept. 1993, pp. 186-189.

[11] H. Akamatsu et al., "A low power data holding circuit with an intermittent power supply scheme for sub-1V MT-CMOS LSIs," in Symp. VLSI Crircuits Dig. Tech. Papers, June 1996, pp. 14-15.

[12] D. Takashima et al., "Stand-by/active mode logic for sub-1V 1G/4Gb DRAMs," in Symp. VLSI Circuits Dig. Tech. Papers, May 1993, pp. 83-84.

[13] T. Kuroda et al., "A 0.9V 150MHz 10mW 4mm^2 2-D discrete cosine transform core processor with variable-threshold-voltage scheme," in ISSCC Dig. Tech. Papers, Feb. 1996, pp. 166-167.

[14] M. Mizuno et al., "Elastic-Vt CMOS circuits for multiple on-chip power control," in ISSCC Dig. Tech. Papers, Feb. 1996, pp. 300-301.

[15] T. Ooishi et al., "A well-synchronized sensing/equalizing method for sub-1.0V operating advanced DRAMs," in Symp. VLSI Circuits Dig. Tech. Papers, May 1993, pp. 81-82.

[16] L. S. Y. Wong and G. A. Rigby, "A 1V CMOS digital circuit with double-gate-driven MOSFET," in ISSCC Dig. Tech. Papers, Feb. 1997, pp. 292-293.

[17] K. Ishibashi et al., "A 1-V TET-load SRAM using a two-step word-voltage method," in ISSCC Dig. Tech.Papers, Feb.1992, pp.206-207.

[18] K. Ishibashi et al., "A 6-ns 4-Mb CMOS SRAM with offset-voltage-insensitive current sense amplifiers," IEEE J. Solid-State Circuits, vol. 30,pp. 480-486, April 1995.

[19] H. Morimura et al., " A 1-V 1-Mb SRAM for portable equipment," in Symp. Low Power Electronics and Design, August 1996, pp. 61-66.

[20] H. Mizuno et al., "Driving source-line (DSL) cell architecture for sub-1-V high-speed low-power applications," in Symp. VLSI Circ. Dig. Tech. Papers, June 1995, pp. 25-26.

[21] H. Yamaguchi et al., "A 0.8V/100MHz/Sub-5mW-operated mega-bit SRAM cell architecture with charge-recycle offset-source driving (OSD) scheme," in Symp. VLSI Circ. Dig. Tech. Papers, June 1996, pp. 126-127.

[22] K. Itoh et al., "A deep sub-V, single power-supply SRAM cell with multi-VT, boosted storage node and dynamic load," in Symp. VLSI Circ. Dig. Tech. Papers, June 1996, pp. 132-133.

[23] Y. Yamaguchi et al., "Features of SOI DRAM's and their potential for low-voltage and/or giga-bit scale DRAM's," IEICE Trans. Electron., vol. E79-C, pp. 772-780, June 1996.

[24] S.Kuge et al., "SOI-DRAM circuit technologies for low power high speed multi-giga scale memories," in Symp. VLSI Circ. Dig. Tech. Papers, 1995, pp. 103-104.

[25] Y. Yamaguchi et al., "Low-voltage operation of high-resistively load SOI SRAM cell by reduced back-gate-bias effect," IEICE Trans. Electron., vol. E78-C, pp. 812-817, July 1995.

[26] K. Ueda et al., "A CAD-compatible SOI/CMOS gate array having body-fixed partially-depleted transistors," in ISSCC Dig. Tech. Papers, Feb. 1997, pp. 288-289.

[27] K. Shimomura et al., "A 1V 46ns 16Mb SOI-DRAM with body control technique," in ISSCC Dig. Tech. Papers, Feb.1997, pp.68-69.

[28] F. Assaderaghi et al., "A novel Silicon-on-insulator (SOI) MOFET for ultralow voltage operation," in 1994 Symp. Low Power Electronics Dig. Tech. Papers, pp. 58-59.

[29] T. Fuse et al., "A 0.5V 200MHz 1-stage 32b ALU using a body bias controlled SOI pass-gate logic, "in ISSCC Dig. Tech. Papers, Feb. 1997, pp. 286-287.

[30] T. Douseki et al., "A 0.5V SIMOX-MTCMOS circuit with 200ps logic gate," in ISSCC Dig. Tech. Papers, Feb. 1996, pp. 84-85.

[31] F. Morishita et al., "Leakage mechanism due to floating body and countermeasure on dynamic retention mode of SOI-DRAM," in Symp. VLSI Tech. Dig. Tech. papers, 1995, pp. 141-142.

[32] T. Tanigawa et al., "Improvement of refresh characteristics by SIMOX technology for giga-bit DRAM's," IEICE Trans. Electron., vol. E79-C, pp. 781-786, June 1996.

[33] S. Tomishima et al., "A long data retention SOI-DRAM with the body refresh function," in Symp. VLSI Circ. Dig. Tech. Papers, June 1996, pp. 198-199.

[34] Y. Tosaka et al., "Theoretical study of alpha-particle-induced soft errors in submicron SOI SRAM," IEICE Trans. Electron., vol. E79-C, pp. 767-771, June 1996.

[35] V. K. De et al., "Random MOSFET parameter fluctuation limits to gigascale integration (GSI)," in Symp. VLSI Tech Dig. Tech. Papers, June 1996, pp. 198-199.

[36] J. D. Meindl et al., "The impact of stochastic dopant and interconnect distributions on gigascale integration," in ISSCC Dig. Tech. Papers, Feb. 1997, pp. 232-233.

Chapter 8

Low-voltage Low-power High-speed I/O Buffers

R. Leung
LSI Logic Corporation; 1501 McCarthy Blvd., E-186, Milpitas CA95035, U.S.A. ; (408)954-4468 (phone), (408)433-7719 (fax), leung@lsil.com

Keywords: high-speed, low-power, buffers

Abstract: High-speed data transport is needed for today's high-performance systems, but power consumption is a major concern, especially for portable equipment. Various popular input/output (I/O) interface schemes will be described, covering typical speed, key circuit design issues, configurations and power consumption. Among popular low-swing low-power solutions, LSI Logic's low-voltage differential signalling (Hyper-LVDS™) buffers will be covered in more details. The list of I/O's discussed include: HSTL, GTL/NTL, PCML, PECL, USB and matched-impedance buffers.

1. INTRODUCTION

High-speed data transfer and low power consumption used to be mutually exclusive design constraints that designers had to trade-off one for the other. However, with the advance of sub-micron silicon technology and rapidly growing demand of portable equipment such as laptop computers, personal digital assistants (PDA's) and wireless communication devices, high performance and low power consumption are required at the same time for competitive system design. While the laws of physics govern that a panacea is impossible, a good understanding of the system characteristics and different I/O schemes will help chip and system designers to appropriately optimize power consumption for the desired performance.

2. WHERE DOES THE POWER GO?

In order to design a low-power system, we have to understand how electrical power is consumed. In the process of data transfer, we can generally define the total power consumption as follows:

$P_{total} = P_{AC} + P_{DC} + P_{overlap} + P_{leakage} + P_{ringing}$

$P_{AC} =$ _ $f C V^2$ where f is the frequency of switching, C is the load capacitor and V is the voltage swing

P_{DC} = DC current consumed in the buffers for biasing, etc.

$P_{overlap}$ = Current consumed as a result of the overlap of "ON" state of the P and N transistors

$P_{leakage}$ = reverse junction and sub-threshold leakage in the buffers

$P_{ringing}$ = power consumed as a result of mismatch between impedance of driver and load

The components in the above equation for a particular system should be individually understood in order to appropriately optimise the power consumption. They will be further discussed in the sections below and will be the criteria for comparison as we look at different kinds of I/O buffer schemes.

2.1 P_{AC}

The frequency of switching is typically set as the design target so there may not be much room for negotiation. However, if data transmission is continuous, switching power can be traded off against other components such as DC power. This ties to the other parameters in this component. The voltage swing is clearly a key factor since the power consumed in switching is proportional to the square of the amplitude. But, in order to maintain a low swing, over variation of operating conditions, a DC biasing scheme is typically needed to achieve the low power consumption at high speed of data transfer. The last parameter, capacitance is always minimised to reduce the capacitive load in a system.

2.2 P_{DC}

The magnitude of this component varies greatly from one type of buffer to another. As mentioned above, in order to minimise the AC switching power, DC power consumption is needed to bias the circuit so that the small voltage swing can be appropriately controlled to accommodate the variations in process conditions, voltage supply and operating junction temperature. Receivers often consume large currents because they need to be kept in saturation to maitain gain. So, the trade-off has to be made between these two components and if the switching pattern of data is well-defined and well-understood, the right optimisation can be made.

2.3 $P_{overlap}$

This is the power consumed during the transition when transistors, which are normally controlled in complementary fashion, are momentarily switched on; leading to a conducting current path. This is most common, and most detrimental, in the output stage of a full CMOS buffer where the large P and N-channel output transistors are simultaneously turned on during transition. This does not only waste power but also can cause signal integrity issues such as ground and power rail voltage bounce.

2.4 $P_{leakage}$

This comes from diffusion junction and MOS transistor sub-threshold leakage. Compared to other factors, this should be relatively minor in a high-speed system. However, if there is a significant power down period or battery life is a concern, this component should be factored in. Since this is primarily process technology dependent and generally not buffer type dependent, it will not covered in subsequent discussions.

2.5 $P_{ringing}$

This particular component is not well-understood as to its impact on power consumption. However, if there is a significant mismatch between driver and receiver, the ringing can cause false transitions which will waste power or cause the receiver to partially turned on even the false transition does not go through a full swing. While most system designers study this area to address timing issues, few would pay attention to its impact on power consumption. Detailed modelling and simulation should be done to ensure no surprises.

3. VARIOUS TYPES OF BUFFERS

The following buffers will be discussed as to their advantages and disadvantages:

CMOS
HSTL

GTL/NTL
PCML
PECL
USB
Matched-impedance Buffer
Hyper-LVDS™

3.1 CMOS

This is the most common type of buffers used in system design today. It suits most digital interface requirements and consumes very little power during standby. The basic concept is to control the On/Off of the two driving transistors in a complementary fashion to charge and discharge the load to V_{dd} and V_{ss} respectively. Most of the issues centre around the imperfect timing of those complementary signals, which will lead to simultaneously switching both the P and N-channel transistors On. Since the output resistance is not well-controlled, it can also create a lot of noise on the supply rails due to the fast-changing current drawn across the package inductors of the supply pins.

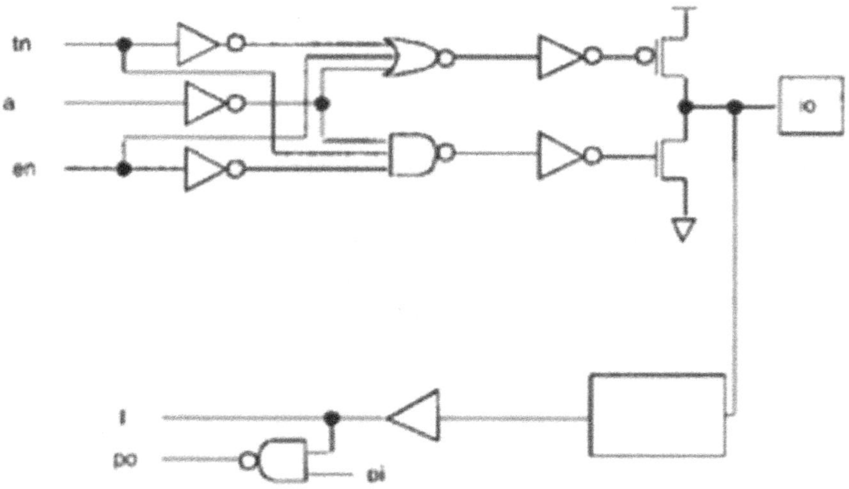

Figure 1. A typical tri-state CMOS buffer.

Its advantages:

- simple and straight forward circuit design

- easily scalable for different drive strength (or output impedance)
- easily scalable for different voltage supply levels
- robust noise margin levels
- less sensitive operating condition variations
- zero P_{DC}
- Its disadvantages:
- high $P_{overlap}$ - causing large ground bounce and other signal integrity issues
- output impedance is not well-controlled, leading to excessive ringing
- P_{AC} significant with increasing frequency

3.2 HSTL (High-Speed Transistor Logic)

Several classes of buffers are available for specific I/O schemes. Intention was to specify pre-determined buffer configurations so as to standardise interface levels. This JEDEC standard will enable higher level of flexibility, technology (CMOS vs. TTL vs. ECL) independent and somewhat voltage independent (since a separate output supply voltage source is applied).

Advantages:

- standardised interface
- different classes for different considerations
- can be used for high-speed interface with good impedance control
- Disadvantages:
- several classes of buffers can be confusing
- static current, Iddq issues
- typically not power optimised

3.3 GTL/NTL (Gunning Transistor Logic / NMOS Transistor Logic)

These two types of buffers are similar in its system and circuit design philosophies. The main difference is the termination voltage for each scheme; where it is 1.2V for GTL and 1.5V for NTL

Advantages:

- low P_{AC} - due to small output voltage swing
- simple circuit structure leads to low latency
- can be used in point-to-point as well as back-plane applications
- can be in single-ended or differential configurations
- output impedance well-controlled

– not sensitive to V_{dd} supply noise

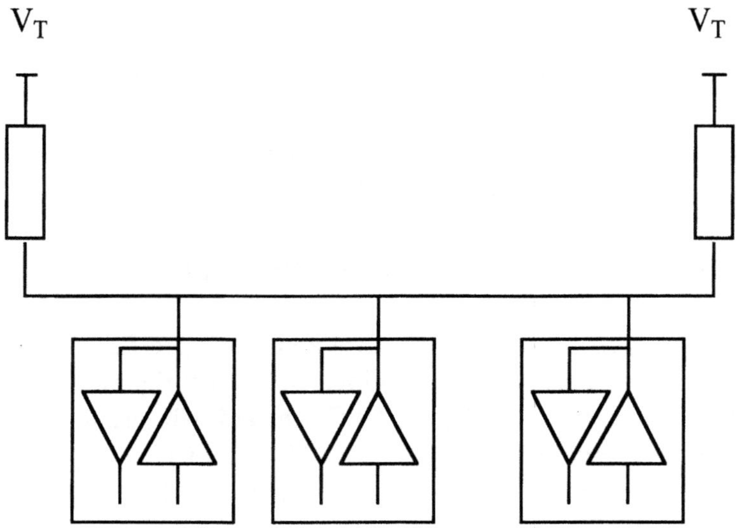

Figure 2. GTL/NTL buffer configuration

Disadvantages:
- consumes DC power
- AC & DC power flows to ground, many ground pins required to maintain signal levels and integrity
- termination voltage required
- high transient current so simultaneously switching noise is a concern

3.4 PCML (Pseudo Current Mode Logic)

PCML uses a current steering scheme whereby a fixed amout of current is steered between 2 complimentary I/O pads to a V_{OH} terminated load

Advantages:
- low simultaneously switching noise
- output impedance is well-controlled
-

Disadvantages:
- a control loop or DC current is required to set the output sink current I_{OL}
- termination voltage needed

3.5 PECL (Pseudo Emitter Coupled Logic)

This can be viewed as the complementary scheme to PCML except PECL uses a current source into a termination resistor (not Vdd referenced)

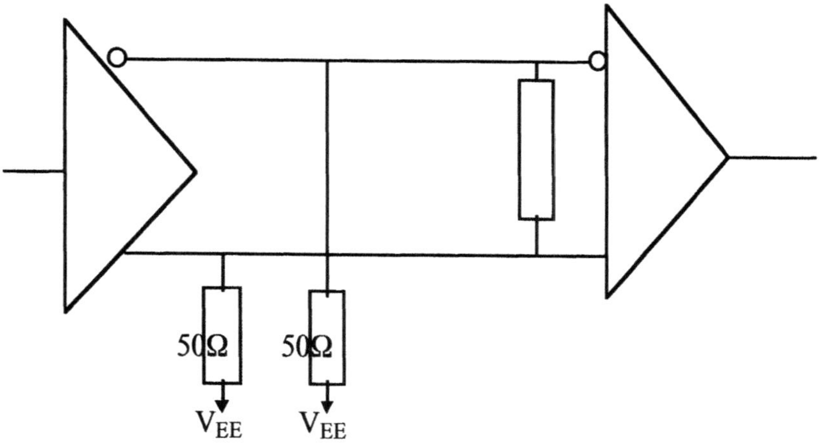

Figure 3. PECL buffer configuration.

Advantages:
- low simultaneously switching noise
- output impedance is well-controlled

Disadvantages:
- DC bias current
- Requires a termination voltage capable of sinking current.
- (most schemes only required termination to source current)
- termination voltage needed
- poor 5V to 3.3V interfacing

3.6 USB (Universal Serial Bus)

This was created for a lot of the consumer oriented applications so the data speed tends to be lower. The main focus is careful slew rate control for ease of interconnection with consumer electronic products. Otherwise, its general operation and signal levels are very similar to CMOS buffers. These I/O buffers are typically in differential mode.

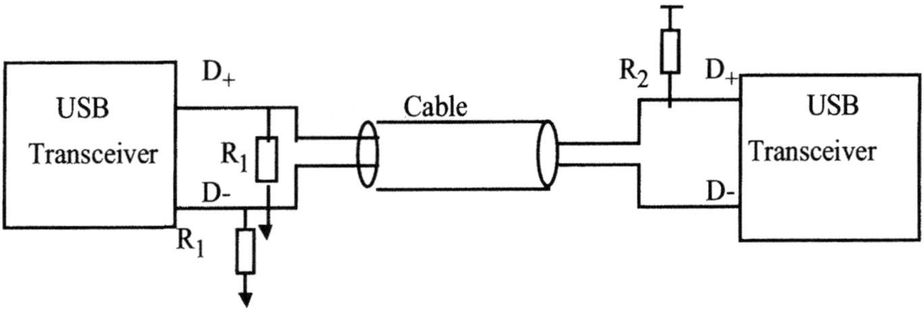

R_1 = 15K Ω, R_2 = 1.5K Ω
Full speed function: Cable = Twisted Pair Shielded, 5 Meters max., 90Ω +/- 15%
Low speed function: Cable = Untwisted/unshielded, 3 Meters max., 90Ω +/- 15%
(R_2 connected to D-)

Figure 4. USB buffer configuration.

Advantages:
- low simultaneously switching noise because of good slew rate control
- less stringent requirement on connection such as cable specifications
- zero DC power

Disadvantages:
- complex circuit design for slew rate control
- slow data throughput

3.7 Matched-Impedance Buffer

Output impedance changes with different process and operating conditions, making hard to match the load. Schemes mentioned above generally use analogue biasing schemes or external resistors to achieve controlled output impedance but at the sacrifice of DC power consumption. This buffer was created by applying digital feedback and control methods to attain the same design goal. The monitor circuit can be shared by all such buffers on the same chip, thus saving valuable silicon area.

Advantages:
- low simultaneously switching noise
- output impedance is well-controlled
- no DC power

Disadvantages:
- relatively more complex circuit design
- chip-wide distribution of control signals may complicate routing issues
- high power for medium performance due to large signal swing

3.8 Hyper-LVDS™ (Low-Voltage Differential Signals)

This buffer is designed to satisfy the IEE Standard for Low-Voltage Differential Signals for SCI. More details can be found in Appendix A.

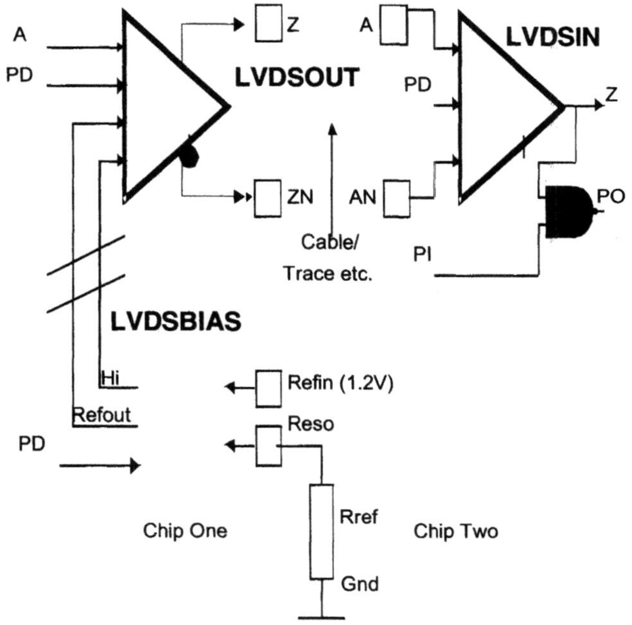

Figure 5.

Advantages:
- low simultaneously switching noise
- output impedance is well-controlled
- self-terminated
- low EMI

Disadvantages:
- complex circuit design
- DC bias current
- reference resistor & voltage required to set bias currents and operating point

4. SUMMARY

Table 1. Comparison of different I/O buffers

Parameters	CMOS	GTL/PCML	PECL	USB	MI	Hyper-LVDS™	NTL
Threshold	_ V_{dd}	0.8V	V_{ss}+1V	V_{dd}-1V	_ V_{dd}	_ V_{dd}	1.2V
Swing	V_{dd}	0.8/1.0V	0.4/0.8V	0.8V	V_{dd}	V_{dd}	0.25V
$freq_{max}$	~100M	~200M	~900M	~250M	12M	~150M	~1000M
$V_{termination}$	none	1.2/1.5V	1.3/1.6/2V	1.6V	none	none	none
Prop Delay	Medium	Fast	Fast	Fast	Slow	Slow	Fast
Noise	Good	Average	Average	Average	Good	Excellent	Margin
P_{DC}	Low	High	High	High	Low	Low	Medium
P_{AC}	High	Low	Low	Low	Low	High	Low
SSO Noise	High	High	High	High	Medium	Medium	Low

5. CONCLUSION

As discussed above, optimising the power consumption of a system requires good understanding of the desired performance, switching behaviour and other application constraints. While simple CMOS buffers can be used for general purpose interface, system designers can fine-tune the power consumption and data throughput trade-off by carefully considering the other buffers.

6. REFERENCES

JEDEC Standard JESD8-A (1994) Interface Standard for Nominal 3V/3.3V Supply Digital Integrated Circuits. Electron Industries Association, Virginia.

JEDEC Standard JESD8-3 (1994) Gunning Transceiver Logic (GTL) Low-Level, High-Speed Interface Standard for Digital Integrated Circuits. Electron Industries Association, Virginia.

EIA/JEDEC Standard EIA/JESD8-6 (1995) High Speed Transceiver Logic (HSTL) A 1.5V Output Buffer Supply Voltage Based Interface Standard for Digital Integrated Circuits. Electron Industries Association, Virginia.

TIA/EIA Standard TIA/EIA-644 (1996) Electrical Characteristics of Low Voltage Differential Signaling (LVDS) Interface Circuits. Telecommunications Industry Association, Virginia.

Universal Serial Bus Specification 1.0 Final Draft Revision, November 13, 1995.

7. BIOGRAPHY

Raymond Leung earned his BSEE and MSEE degrees from Columbia University, New York, in 1981 and Stanford University, California, 1982 respectively. He was senior design engineer with Philips Semiconductor (82-86, 86-89) and with Silicon Compilers Inc. (86-87). He has managed the cell library, memory and I/O buffer development groups at LSI Logic Corporation where he is currently the Director of Mixed-Signal Development.

Kiyoo Itoh received Ph.Degree from Tohoku University, Japan in 1976. He is now the Senior Chief Scientist of Hitachi Central Research Laboratory, Japan. He was a visiting MacKay Lecturer at the University of California at Barkeley in 1994, and a Visiting Professor at the University of Waterloo, Canada in 1995. As a result of his 25 year DRAM career in Hitachi, he hold over 120 patens including the folded data (or bit) -line circuit for DRAMs. In addition, he and his team have led the DRAM technology development through over 100 papers and presentations. He is the author of VLSI Memory Design (Baifukan, in Japanese), published in 1994. Dr. Itoh received numerous awards including the 1993 IEEE Solid-State Circuits Award, and the IEEE Fellow Award.

Chapter 9

Microelectronics toward 2010

T. Yanagawa[1], *S. Bampi*[2], G. Wirth[3]

[1] *General Secretary - The Information Processing Society of Japan Shibaura-Maekawa Bldg. 7th Floor 3-16-20 Shibaura, Minato-ku, Tokyo 108-0023, JAPAN; Tel: +81-3-5484-3535, Fax: +81-3-5484-3534 E-mail: yanagawa@ipsj.or.jp*

[2] *Prof. at the Informatics Institute and Electrical Engineering Department UFRGS – Federal Univ. of Rio Grande do Sul; P.O. Box 15064 – 91501-970 Porto Alegre, BRAZIL Tel: +55-51-316-6812, Fax: +55-51-319-1576; E-mail: bampi@inf.ufrgs.br*

[3] *Prof. at the Electrical Engineering Department and Informatics Institute UFRGS – Federal Univ. of Rio Grande do Sul; P.O. Box 15064 – 91501-970 Porto Alegre, BRAZIL; Tel: +55-51-316-6828, Fax: +55-51-319-1576; E-mail: wirth@inf.ufrgs.br*

Key words: Microelectronics, Moore's law, roadmap, limitations, breakthroughs

Abstract: Middle-term perspectives of process-device technology of semiconductor integrated circuits are described. Even though Moore's law is generally considered to hold good for 10-15 years more, many of the current technologies are foreseen to face growth limitations and thus undergo innovative changes. Problems and possible breakthroughs are discussed for lithography, transistor size, interconnections and power dissipation as the principal factors of such limitations. Future directions to expand functionality and performance of integrated circuits are also described.

1. INTRODUCTION

Microelectronics is one of the most marvelous technologies developed in this century. It has been playing a prominent role in raising the electronic industry to the leading position and will continue to be a keystone to accelerate the development and integration of computers, communications and consumers which are the three core constituents for the emerging advanced information society. The key components here are semiconductor integrated circuits. They came into being as a result of the effort to assemble smaller and lighter circuits. However their benefits to electronic equipment are not confined to this original aim but extend to wider aspects such as low cost, high performance and high quality. These miraculous effects are of character that is enhanced as the number of components on a semiconductor chip increases. Therefore this number or, in other words, the *integration level* is the universal index to show the progress of integrated circuit technology.

Integrated circuits integrate components by virtue of relying on various technologies and methodologies ranging from semiconductor physics, materials science, electronics, computer science, to system architectures. Specifically process-device technology and design technology are center pieces without which the integrated circuit business cannot exist. The former technology defines the upper bound of realizable performance and functions while the latter gives specific functionality to products. Process-device technology (expressed as semiconductor technology or just technology hereafter) is important since it is the engine for pushing forward the integration level. It is the core competence for semiconductor manufacturers even in the market driven circumstances.

In this chapter a middle term outlook for semiconductor technology is presented, targeting at 2010. An accurate prediction is not easy because of the rapid technology progress in this area but this is a challenging task since the progress of integrated circuits is closely correlated with that of information society. In the following sections, general trends of technology progress are overviewed at first and then problems and solutions to follow the trends are discussed for major elementary technologies. Lastly some directions to further advance the performance and functionality of future integrated circuits are described.

2. TRENDS OF SEMICONDUCTOR TECHNOLOGY

When discussing the prediction of semiconductor technology, there are two guidelines which should be taken into account: Moore's law (Schaller, 1997) and the SIA semiconductor roadmap (ITRS, 1999). Moore's law was publicized originally in 1965 by Gordon E. Moore, one of the founders of Intel Corporation (Moore, 1965). He projected that the integration level would grow at the rate of 1000 times per decade. Later in 1975 the growth rate was mitigated to 100 times per decade or 2 times in 18 months (Moore, 1975). As shown in Figure 1, this law has guessed right for almost a quarter century the capacity growth of DRAMs which have been leading the progress of semiconductor technology from 1970 to 1990, as a technology driver. Minimum feature size, which is the principal factor to determine the integration level, is about the same for DRAMs and microprocessors but the growth of the number of transistors in microprocessor has been somewhat slower than that of DRAM capacity. This is because more complex interconnections of microprocessors reduce the transistor density. In the 1990's, microprocessor performance has driven the technology, such that transistor channel lengths have been reduced aggressively, beyond the technology node figure of merit.

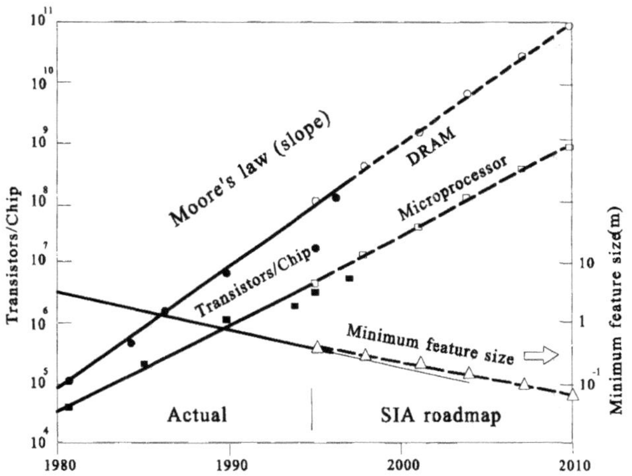

Figure 1. Growth of integration level in the past and future: Moore's law and the semiconductor roadmap.

One can interpret that Moore's law has proved right because it has become the common development target throughout the integrated circuit industry. It has made people believe that the law is attainable and that one company falls behind competitors when its technology cannot follow it. This power would not have lasted long if their target could not be attained no matter how hard they strove or, on the contrary, could be surpassed with ease. The greatness of Moore's law comes from the clear insight into the engineering capability.

The technology roadmap for semiconductors (expressed as roadmap hereafter) provides a 15-year outlook for the growth path of integration level for major integrated circuits products and also of the detailed process, device and design technology. It was published by Semiconductor Industry Association (SIA) of USA in 1992 (SIA, 1992) and was revised twice: in 1994 and 1997. In 1999 the document was again revised, with inputs from all regions of the world, and named International Technology Roadmap for Semiconductors (ITRS, 1999). Its abstract is shown in Figure 1 and Table 1. This roadmap is widely accepted and thus there is no point in raising different views except for some minor points. The roadmap is based on the assumption that Moore's law will be effective until at least 2014 and its plot over Moore's law shows the straight extension as shown in Figure 1. As the name "roadmap" suggests, its primary object is to provide researchers and engineers with a common target so that their efforts are brought together to solve more and more difficult technical problems. Accordingly the roadmap is to be subject to periodic review.

Table 1. Abstract of the International Roadmap for Semiconductors (ITRS, 1999): Performance of Packaged Chips.

Year	*1999*	*2002*	*2005*	*2008*	*2011*
Technology Node	*180nm*	*130nm*	*100nm*	*70nm*	*50nm*
Clock frequency, On chip, High-performance (MHz)	1200	1600	2000	2500	3000
Number of metal layers.	6-7	7-8	8-9	9	10
Power dissipation (W)	90	130	160	170	174
Power supply (V)	1.8	1.5	1.2	0.9	0.6

3. PERSPECTIVES OF KEY TECHNOLOGIES

Progress of technology generally follows the following steps:

1. Improvement on current technology
2. Replacement with new technology.

It is desirable for a technology to have a large room for improvement because this lengthens its life and enables the accumulation of technical assets in the meantime. This viewpoint is important when selecting a new technology. Fortunately integrated circuits have been blessed with this sort of technologies such as silicon for material, light for lithography and MOS transistors for active devices. The remarkable progress of integrated circuits can be attributed to this coincidence. However, there are limitations even for such technologies. When a technology approaches its limitation, its improvement tends to become more and more costly and time consuming and a chance for changeover to a new technology increases. As shown in Figure 2, the changeover occurs when the incremental cost for improvement exceeds the total development cost of a new technology with higher capability (step 2). Four key factors in the development of integrated circuits are analyzed with regard to their limits and possibilities of surmounting these lits are discussed below.

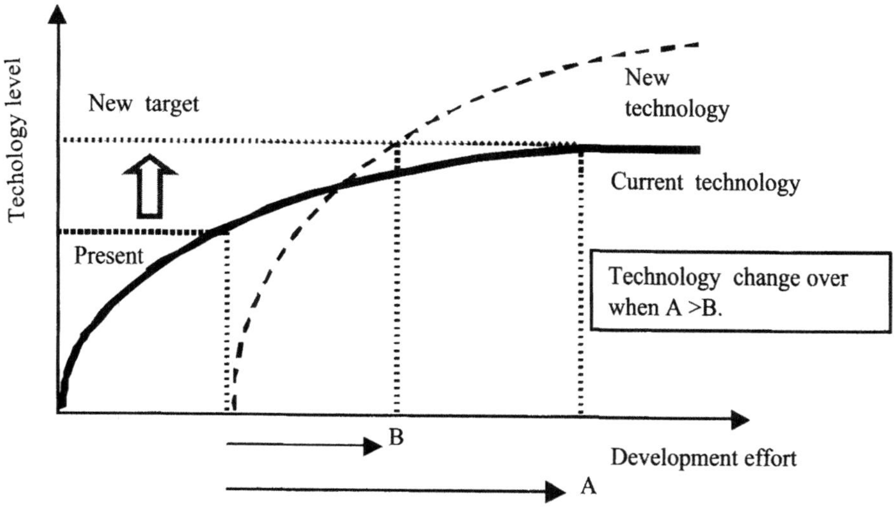

Figure 2. General tendency of technology changeover.

Photolithography

High productivity and thus low cost of integrated circuits have been brought about by the batch production technique. Lithography whose role is to define horizontal dimensions of transistors and interconnections is one of the important processes which enable the batch production. As most integrated circuits nowadays take two-dimensional structures using only the surface layers of semiconductor substrates, lithography is the key to increase the integration level. Photolithography or optical lithography in which a light beam is used to expose light sensitive resin films is the most widely used lithography technique today. The light beam is very convenient to handle because it can be focused by lenses and its path can be bent by mirrors and prisms in the air. For this reason the light beam has been used for more than 30 years despite the challenges of other energy beams with potentially higher capability, in particular higher linewidth resolution.

An image resolution through a lens of projection exposure equipment is proportional to the wavelength of light. Therefore shorter wavelengths have been used as the shrinkage of pattern size has been done. Table 2 shows this relation. For the 0.13μm photolithography to be applied to 2Gbit DRAM, the use of ArF laser with a wavelength of 0.193μm is necessary. However the following difficulties emerge when the wavelength reaches to this level and lower:

A light source with high intensity becomes difficult to obtain.

Absorption of light through lenses and photomasks increases at shorter wavelengths.

Absorption of light in the photoresist layer also increases and makes it difficult for the light to reach the bottom of thick photoresist films.

The depth of focus of the lens decreases in proportion to larger numerical aperture and shorter wavelength, and uneven wafer surfaces deteriorate the pattern resolution capability across the wafer.

Further development of next generation lithography, an industry-wide effort, envision the possibility of extreme UV (EUV) light to be used for lithography down to 0.07 μm minimum on-wafer features. Subwavelength lithography techniques, with 157 nm EUV source, coupled to phase-shift masking and high numerical aperture optics, has been developed to extend the applicability of optical lithography. Light with a 157 nm wavelength is absorbed by organic materials and oxygen, so new resist materials and oxygen-free exposure tools have to be developed. This could be about the limit of photolithography. Fortunately there already exist alternatives to light: electron beam projection, single beam direct write, ion beam projection and proximity X-ray lithography. Some of them are being used practically at present. They have a research history of more than 20 years and there is a considerable reserve of technical assets. However there will be a large impact on manufacturing when the light is replaced with another energy source for the first time in the history of integrated circuits manufacturing.

Table 2. Minimum feature size and light sources of photolithography

Design rule (μm)	*DRAM generation (Mbits)*	*Light source*	*Wave length (μm)*
0.5	16	*i*-line (Hg)	0.365
0.35	64	*i*-line (Hg)*	0.365
		or KrF laser	0.248
0.25	256	KrF laser	0.248
0.18	1000	KrF laser	0.248
0.13	2000	KrF* laser	0.248
		or ArF laser	0.193
0.10	8000	ArF* laser	0.193
		or F_2 laser	0.157
0.07	16000	F_2* laser	0.157
	64000	Not Available	N/A

* with resolution enhancement

MOS transistor

- The transistor size, especially the channel length, has been following the course of shrinkage in order for both increasing the integration level and reducing the switching energy per logic transition, which is the product of switching delay and power dissipation. This tendency is plotted in Figure 3. However the functions of transistor as an active device are disturbed by the following phenomena when the shrinkage progresses:
- The short channel effect reduces the threshold voltage and increases the off-state leakage current. Furthermore, there is a lower limit on channel length, because too thin depletion regions are subject to quantum mechanical tunneling of charge carriers from source to drain, leading to high leakage currents
- Enough area for heat removal through substrate cannot be secured. This is a problem for any densely packed integrated circuit, and will limit the device density, since overheating can cause malfunction.
- Variations of transistor characteristics due to fluctuations of process parameters become relatively large, and the lowering of supply voltages, mandatory to face the imperative reduction of CMOS power dissipation, makes the circuits more sensitive to these fluctuations. Many of these fluctuations arise from processes that are stochastic in nature and can not be improved by lithography, like ion implantation. For a doping concentration $N_A = 5x10^{18}$ cm^{-3} there will be less than 50 dopants in the channel region of a transistor with W=L=50 nm. The actual number of dopants found in the channel follows a normal distribution. This leads to significant scattering in threshold voltage and drive current (Asenov, 1998). Even if the number of dopants in the channel could be precisely controlled, there would still be electrical parameter variations due to the microscopic arrangement of the dopants. To solve the problem, alternatives to ion implantation are necessary to position individual dopant atoms in specific lattice sites.
- The leakage current through the gate oxide increases exponentially as oxide thickness decreases. Quantum mechanical tunneling currents exclude the use of SiO_2 for layers below 2 nm, needed for devices with L<100nm (ITRS, 1999). The application of CMOS compatible high dielectric constant (high κ) films to replace silicon dioxide is being developed for technologies below 0.10 μm. One benefit of high κ materials is the possibility of using thicker films in order to decrease gate leakage. However, dielectric film thickness of the order of the channel length results in increased fringing fields from the gate to source/drain regions and compromised short channel performance(Cheng, 1999).
- The channel length has been scaled down faster than the supply voltage, leading to increasing electric fields in the device. High fields near the

drain can cause avalanche breakdown by ionizing a large number of semiconductor atoms (generation of electron-hole pairs), thus causing spurious current and possible device damage. Judicious limits on supply voltage can keep the reliability requirement under control.

- As the channel area and supply voltage decrease the amount of charge required to excite the device scales down. This makes the device more sensitive to soft errors, caused by charge collected from alpha-particle hits. In addition, the signal to noise ratio degrades. Smaller channel areas and lower supply voltages mean less electrons in the channel, and current fluctuations due to fast interface states and oxide traps are enhanced (wirth, 1999). Reduction in defect density to achieve reliable operation of analog, mixed signal and RF applications is a difficult challenge.

For these reasons the practical limit of effective channel length was estimated to be about 0.1μm (Sugano, 1992). This corresponds to the target of 2005 in the roadmap revised in 1999 and there are solid evidences that it was surpassed by aggressive polisilicon line width scaling even in the 0.18μm technology node (Iwai, 1999). The attainment of the roadmap target of 0.07μm in 2008 requires some technology breakthroughs. It should be noted, however, that practical limits are usually discussed based upon certain assumptions which are considered reasonable at a given time. If such assumptions are overthrown by an emergence of new theory or technologies, there can be further progress. This is the usual process which has gone through so far and thus the constant advancement of integrated circuit technology has been realized. Experimentally speaking, transistors with channel length of less than 0.05μm have been fabricated. The past experience tells that the experimental transistors can be transferred to mass production in about 15 years (See Figure 3.). Therefore 0.07μm transistors for commercial products in 2008 will necessarily be achieved, even though concrete solutions are not visible yet for mass production of a fully scaled 0.07-0.05 μm technology node, but cost may become a major issue in scaling down. If scaling down increases the cost too much, it may not be affordable to build a facility to run a 0.05 μm process, although technically possible.

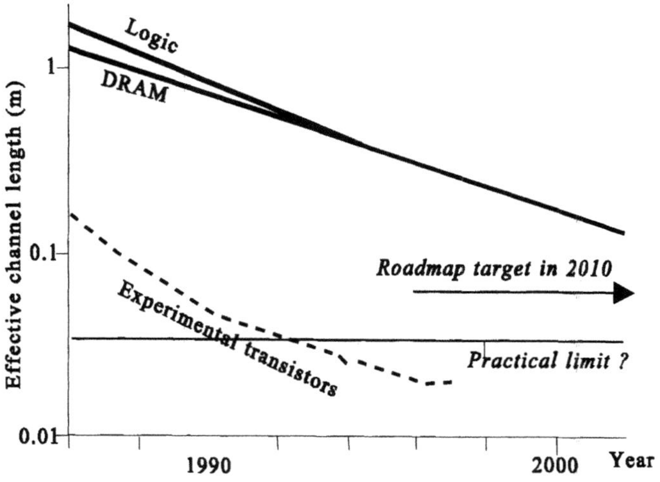

Figure 3. Trends of transistor size shrinkage.

Interconnections

The original role of interconnections is to keep different nodes in a circuit at an equal electric potential. They used to be counted not as circuit components but be transparent like air from the point of circuit performance. They are becoming finer and denser according to the increase of integration level. On the other hand an operating speed required of a circuit is ever increasing. Under these circumstances interconnections are not innocent substances anymore but pose the following profond problems. They can be even a limiting factor to the integration level.

- Wiring delay occupies the dominant part of signal delay between input and output. This case occurs at the design rule below about 0.5μm for CMOS circuits. In other words, efforts to shrink transistor size described in the previous item become meaningless unless the problem of wiring delay can be solved. As shown in Figure 4, the equivalent circuit of a piece of wire becomes more and more complex which in turn makes the design of advanced circuits prohibitively complex. The total elimination of wiring delay cannot be hoped for and thus it is essential to obtain a design method for realizing a circuit with desired characteristics under the existence of wiring delay. This requires the proper management of interactions between electrical and physical designs.

  Chip areas needed to accommodate interconnections reduce the transistor density and hence the integration level as evidenced by the case of microprocessors in Figure 1. One solution is to increase the number of metal layers. Figure 5 shows the trends in the past. It is to be noted that interconnections dominate chip areas as well as fabrication steps. With the adoption of surface flattening processes such as chemical mechanical polishing (CMP), the attainment of 9-10 layer metallization in 2010 will not be an unrealistic target.

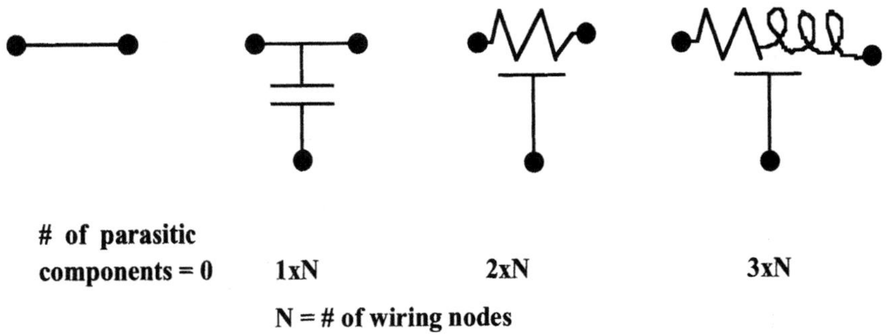

Figure 4. Equivalent circuits for wires.

- Interconnections consume power and their design require specific low power design techniques.

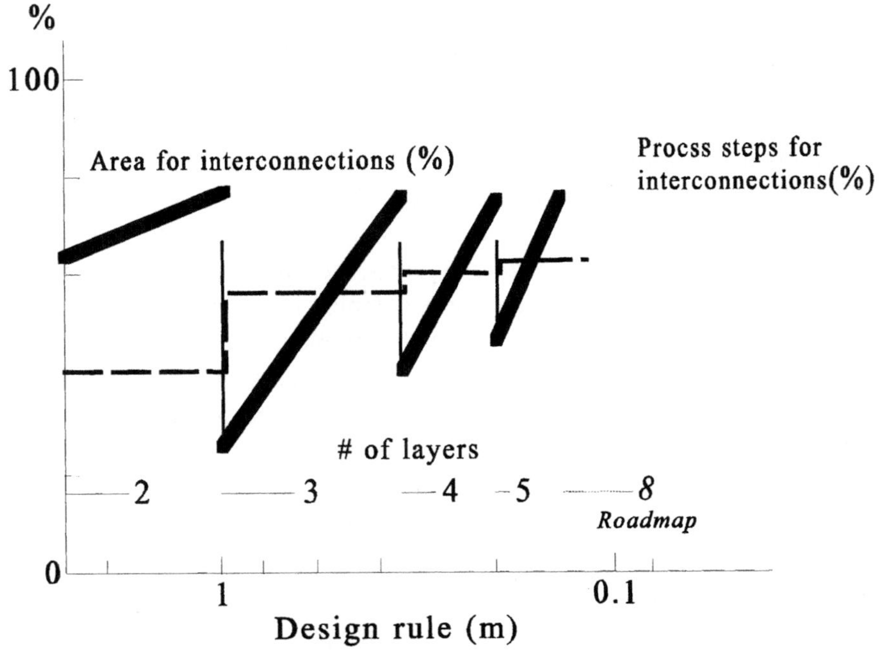

Figure 5. Trends of multilayer metallization.

- As the clock frequency is increased beyond 1 GHz the wavelength becomes comparable to the length of the longest interconnect metal wires, and they do start radiating like antennas. Then the metal wires have to be treated as transmission lines. Beyond the energy radiation problem, to model interconnects as transmission lines is also a challenge for design and CAD engineers.
- As the supply voltage is scaled down, cross-talk becomes an issue for clock and signal wiring lines. Signal integrity rules must be applied to the design.
- Increase of current density causes electromigration and deteriorates the chip reliability. This is the law of nature, and heavier atoms (Cu, instead of Al) are better materials.

New technologies to face the above problems are being developed. Copper metallization was introduced in production in 1998 and new insulator materials with low dielectric constant (low κ) are being studied. Copper is less prone to electromigration and has lower resistivity, 2.2 μΩ/cm, as compared to Aluminium, with 3.3 μΩ/cm.

Interconnect delay depends on the RC constant, and the dielectric constant affects the capacitance and cross-talk directly. Silicon dioxide, nowadays the most commonly used insulator, has an effective dielectric constant of 3.9. Materials under investigation still have to be proven in production, potentially with relative dielectric constant down to 1.5.

But traditional scaling like thinning copper metallization combined with material innovation will not be able to satisfy the long term performance requirements. To overcome the interconnect performance limitations new design paradigms and technology innovations are mandatory. Some of the breakthrough technologies for the above-mentioned interconnection problems are:

Delay insensitive circuits (asynchronous circuits, or globally asynchronous and locally synchronous designs).

On-chip optical interconnections.

Multiple valued logic circuits, or multi-valued bus signaling.

Only delay insensitive circuits can cope with the fundamental limit on signal velocity. Signals travel slower than the velocity of light in vacuum, which requires a field wave approach to design for gigahertz clock frequencies. A front wave in vacuum wil travel only 3 cm during the period of a 10 GHz synchronization signal. For example, to get the response from the processors cache within 1 cycle, none of the memory elements can be more than 1.5 cm away from the central processor. This dimension is comparable to the chip size, and the metal lines connecting the processor to the memory elements are not straight lines. Actually, signals travel much slower, due to higher relative dielectric constants, parasitic capacitances and resistances of metal lines.

The problem of getting the signals in/out of the chips in the GHz frequency range is a great challenge, too. It will be necessary to treat the overall system as a unit, instead of designing parts separately.

Power dissipation

Circuits require energy when in operation, which in turn presents the following two issues:

Limited battery life which is especially important for portable equipment

Generation of heat.

To prolong the battery life, power management is an effective solution as well as the improvement of battery power density. However the full integration capability as projected in the roadmap cannot be utilized for portable equipment because of the limitation of battery life and also heat generation as described below.

Consumed energy generates heat. Semiconductor pn-junctions lose their function when their temperature exceeds a certain limit. These are material constrains that limit the maximum integration level. The problem caused by these effects is serious in the consumer and portable products where cooling methods are restricted practically. The maximum integration level N_{MAX} can be estimated by the following equation:

$$N_{MAX} = 2\Delta T_{MAX}\, t_{pd} / [\, V_{DD}\, R_{pk}\, a\, (C_m + C_w)(V_{DD} - V_T)\,]$$

where, ΔT_{MAX} = maximum allowable temperature increase

t_{pd} = signal propagation delay

R_{pk} = thermal resistance of package

a = gate activity ratio

C_m = transistor output capacitance

C_w = wiring capacitance

V_{DD} = power supply voltage

V_T = threshold voltage

When certain realistic values are substituted in these parameters, the maximum circuit complexity becomes about 30 million gates in the case of air cooling. This is a huge number but corresponds only to the roadmap target in 2004. There are no workable solutions foreseen at present to exceed this limit. This limitation to integration, derived from power dissipation, lead to a *power crisis*, to be dealt with at the system design requirements.

4. BREAKTHROUGHS FOR THE FUTURE DEVELOPMENT

Moore's law uses the number of components in a chip as the vertical scale. However the definition that this number is identical to the integration level is causing the difficulties in technology development as described above. What customers pay for is not the number of transistors and other components in a chip but its function and performance. Therefore it will be meaningful to return to this basic concept. The followings are two directions worthy of attention.

4.1 Reduction of the number of transistors per function

This direction to realize a required function with as small number of transistors as possible is not new. For example, this was the very important issue for memory designers until the invention of one transistor DRAM cell in 1973. This thought will need to be revisited. It is necessary to include parasitics when considering the effective number of components. The followings are some circuit technologies in line with this principle:

- Analog circuit technology
- Multiple valued logic technology
- Delay insensitive circuit technology (for parasitics)
- New functional devices including neuro-devices.

The last item is the hot research subject but it will be out of the timeframe of this paper that those devices come into a practical use widely. The rest are the technologies with long history. The multiplication of logic values is now commercialized for flash memories. Figure 6 shows the prototype of 4Gbit DRAM using two-bit cells (Murotani, 1997). The chip size has been almost halved in comparison with the one using one-bit cells and the same design rule (0.15μm). For logic circuits the compact configuration of gates is a difficult problem. There is a research report, however, that the chip size and power dissipation of multiplier have been reduced to 50% taking advantage of the inherent parallelism of multiple valued logic algorithm (Kameyama, 1988). Additionally the multiple valued logic can be a solution for the interconnection problems as shown in Figure 7 (Kameyama, 1992).

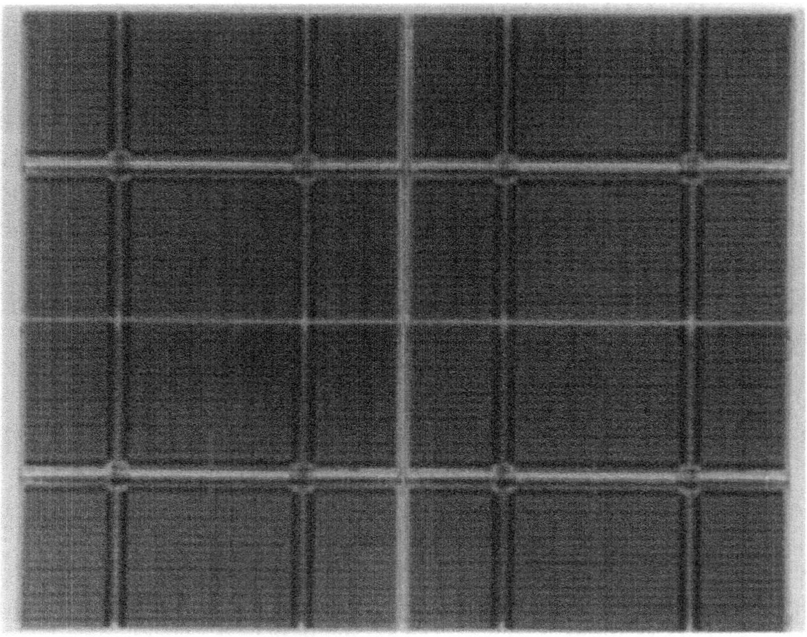

(a)

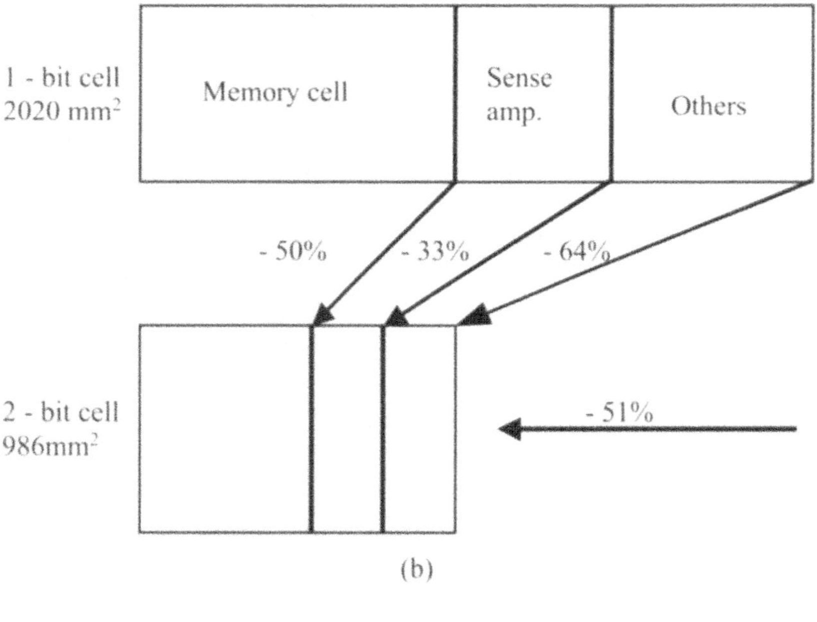

(b)

Figure 6. . 4Gbit DRAM using 2-bit memory cell: (a) chip photomicrograph (33.9x29.0mm^2, 0.15μm CMOS) and (b) comparison of chip size between 2-bit and 1-bit memory cells.

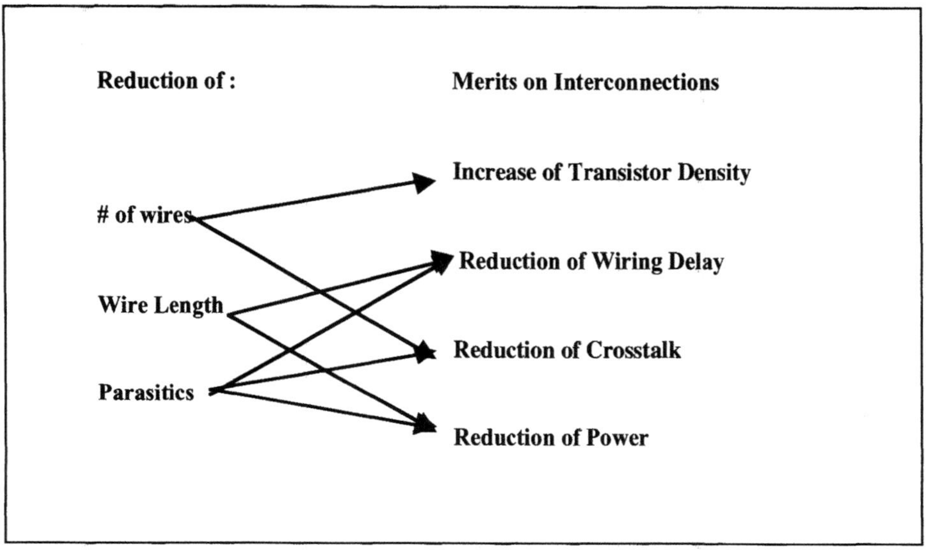

Figure 7. Merits of multiple valued logic on interconnections.

4.2 Use of flexible circuits for high performance

Human brain is believed to have the most advanced computing capability. It has been the familiar discussion subject that how close computers can approach the brain. Gates in integrated circuits operate several orders of magnitude faster than nerve cells. Even in terms of complexity, a chip in 2010 is predicted to contain gates as numerous as more than one hundredth of the number of human nerve cells. However the difference in performance between them will remain extraordinary except for limited computing tasks. This vast difference comes from the difference in numerousness and flexibility of interconnections between computing elements. Neural networks and biological information processing are being researched to pursue an improvement of machine performance along this line, whose practical use will need some more time though. Field programmable gate arrays (FPGA) which are already commercialized have a large flexibility in circuit construction. The use of this feature can be a substitute to approach the high performance of brain. The followings are two of such applications.

Hardware realization of algorithms

Current FPGAs are slower by about an order of magnitude than mask gate arrays. However, hardwired circuits have an inherent advantage in speed by several orders of magnitude over software solutions on general purpose processors. With the use of efficient design tools FPGAs can realize a special purpose hardware quickly and at low cost.

Configurable computing

Within the brain, connections between nerve cells are made through synapse which can change their connectivity according to what has been learned earlier. This is in a sense the adaptive optimization of hardware configurations. The similar function can be performed by FPGAs. This is the research topics named *configurable computing*. Today's FPGAs take millisecond order of time for reconfiguration. If this time can be reduced to microsecond level, on-the-fly circuit modifications become possible for many applications. Needless to say hardware optimization must be performed at the same high speed too. This is a new design paradigm. There have been many research results reported so far and Figure 8 shows one of them. This is the fault tolerant FPGA with 8×8 logic cells each of which is equipped with built-in-self-test capability (Shibayama, 1997). The function of each cell is monitored all the time during operation by the control circuits placed in the chip periphery. When an error is detected in a cell, it is replaced with an unused normal cell. The chip is fabricated by 0.35μm CMOS process and operates at 20MHz.

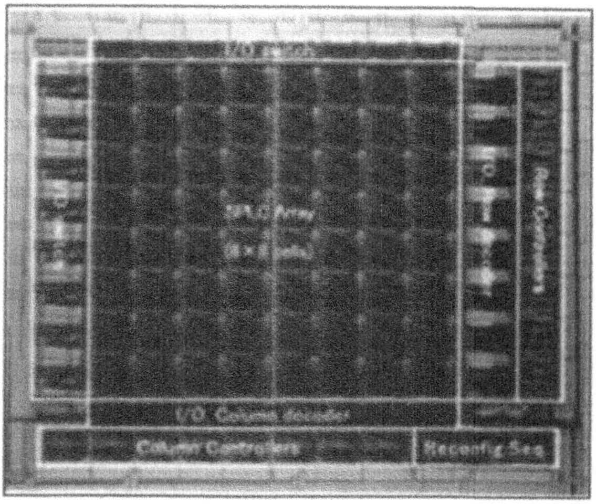

Figure 8.. Chip photomicrograph of 8x8 fault tolerant gate array (1.9x1.6mm^2, 0.35μm CMOS, three layer metal).

5. CONCLUDING REMARKS

The steady progress of technology cannot be attained only by the intelligence of scientists and engineers. It is indispensable to secure financial resources to sustain the development of more and more difficult technology. One important point is which device can be a technology driver in the future. The technology driver is the device which (1) uses the most advanced technology and (2) raises the largest earnings or, in other words, is consumed in a large quantity. It is not easy to find devices to meet these conditions. For example, servers require the highest performance microprocessors but their use will not be so numerous as that of client side equipment. High performance microprocessors are likely the technology driver today, possibly being replaced by mobile equipment ICs before 2010. Another point is to establish a mechanism to alleviate the burden for each technology developer. Partnership and standardization are two important means. The competitive business environment of microelectronics has been the source of the rapid progress of technology and will never subside. Therefore compatibility of cooperation and competition is an important business model issue to be considered in the future.

6. REFERENCES

Asenov, A (1998) Random Dopant Induced Threshold Voltage Lowering and Fluctuations in Sub-0.1 μm MOSFETs: A 3-D Atomistic Simulation Study. *IEEE Trans. on Electron Devices,* **45**, 2505-2513.

Cheng, B et al. (1999) Design Considerations of High-k Gate Dielectrics for sub-0.1μm MOSFETs . *IEEE Trans. on Electron Devices,* **46**, 261-263.

Iwai, H (1999) CMOS Technology – Year 2010 and Beyond. *IEEE Journal of Solid-State Circuits*, **34**, 357-366.

Kameyama, M et al. (1988) A multiplier chip with multiple-valued bidirectional current mode circuits. *IEEE Computer*, **21**, 43-56.

Kameyama, M et al. (1992) Post-binary VLSI system: Interconnection-free computing system. *Journal of the Institute of Electronics, Information and Communication Engineers (Japanese)*, **75**, 400-406.

Moore, G.E. (1965) Cramming more components onto integrated circuits. *Electronics Magazine*, **38**, 114-117.

Moore, G.E. (1975) Progress in digital integrated electronics. *Proceedings of the 1975 IEDM*, 11.

Murotani, T. et al. (1997) A 4-level storage 4Gb DRAM. *Digest of the 1997 IEEE International Solid State Circuits Conference*, 74-75.

Schaller, R.R. (1997) Moore's law: Past, present and future. *IEEE Spectrum*, **34**, 53-59.

Shibayama, A. (1997) An autonomous reconfigurable cell array for fault-tolerant LSIs. *Digest of the 1997 IEEE International Solid State Circuits Conference*, 230-231, 261.

SIA (1994, 1997) *The national technology roadmap for semiconductors*, SEMATECH Inc., Austin TX.

ITRS (1999). *International Technology Roadmap for Semiconductors, 1999 Edition.* SEMATECH Inc., Austin TX.

Sugano, T. (1992) Physical limitations for device performance. *Journal of the Institute of Electronics, Information and Communication Engineers (Japanese)*, **75**, 326-332.

Wirth, G (1999). Mesoscopic Transport Phenomena in Ultrashort Channel MOSFETs. *Solid State Electronics*, **43**, 1245-1250.

7. BIOGRAPHY

Takayuki Yanagawa is currently the General Secretary of the Information Processing Society of Japan. Until 2000 he was Chief Engineer, Office of Corporate Technology Strategies, NEC Corporation. He received BSEE and Dr. Eng. degrees in 1961 and 1974 from the University of Tokyo, respectively. He joined NEC Corporation in 1961 and has been engaged in the development of silicon microwave transistors, bipolar integrated circuits and CAD tools for LSIs. He is a member of IEEE, IEICE and IPSJ.

Sergio Bampi is professor of applied informatics and Microelectronics at the Federal University of Rio Grande do Sul (UFRGS), Porto Alegre, Brazil. He received the B.Sc. degree in Electrical Engineering and Physics from UFRGS in 1979 and the Ms.E.E. and Ph.D. degrees from Stanford University in 1982 and 1987, respectively. He works on MOS devices, circuits and systems design as well as on computer aided design tools. He is a member of IEEE.

Gilson I Wirth was born in Nova Hartz, Brazil, on January 31, 1966. He received the B.Sc. degree in Electrical Engineering and M.Sc. degree in Computer Science from the Federal University of Rio Grande do Sul (UFRGS), Porto Alegre, Brazil, in 1990 and 1994, respectively. In 1999 he received the Dr.-Ing. degree in Electrical Engineering from the University of Dortmund, Germany.

From 1990 to 1992 he was with Embramic Ltda., Porto Alegre, Brazil, where he was engaged in embedded systems design. Since 1999 he is with the Informatics Institute and Electrical Engineering Department at UFRGS, Porto Alegre, Brazil, where he works as a Professor. His research interest include semiconductor device modeling, characterization, noise and scaling problems in MOS device.

Index of Authors

GPSR Compliance

The European Union's (EU) General Product Safety Regulation (GPSR) is a set of rules that requires consumer products to be safe and our obligations to ensure this.

If you have any concerns about our products, you can contact us on ProductSafety@springernature.com

In case Publisher is established outside the EU, the EU authorized representative is:

Springer Nature Customer Service Center GmbH
Europaplatz 3
69115 Heidelberg, Germany

Batch number: 10732314

Printed by Printforce, the Netherlands